I0458458

BENNY THE LOVABLE:
LEARN TIME WITH MAZES FOR KIDS AGES 4-8

COPYRIGHT 2025 AL TRAN

ALL RIGHTS RESERVED. NO PORTIONS OF THIS BOOK
MAY BE REPRODUCED IN ANY FORM WITHOUT
PERMISSION FROM THE PUBLISHER, EXCEPT AS
PERMITTED BY U.S. COPYRIGHT LAW. FOR PERMISSION
CONTACT: AL TRAN AT ALTRAN23@YAHOO.COM

ISBN: 978-1-959376-17-0

PRINTED IN THE UNITED STATES

VISIT OUR AMAZON AUTHOR PAGE TO DISCOVER MORE
OF 'BENNY THE LOVABLE' AND OTHER GREAT BOOKS.

SCAN FOR MORE BOOKS

BENNY THE LOVABLE:
LEARN TIME WITH MAZES FOR KIDS AGES 4-8

By AL Tran

If you enjoy this book,
please give us a review.
We would really appreciate it.

Instruction

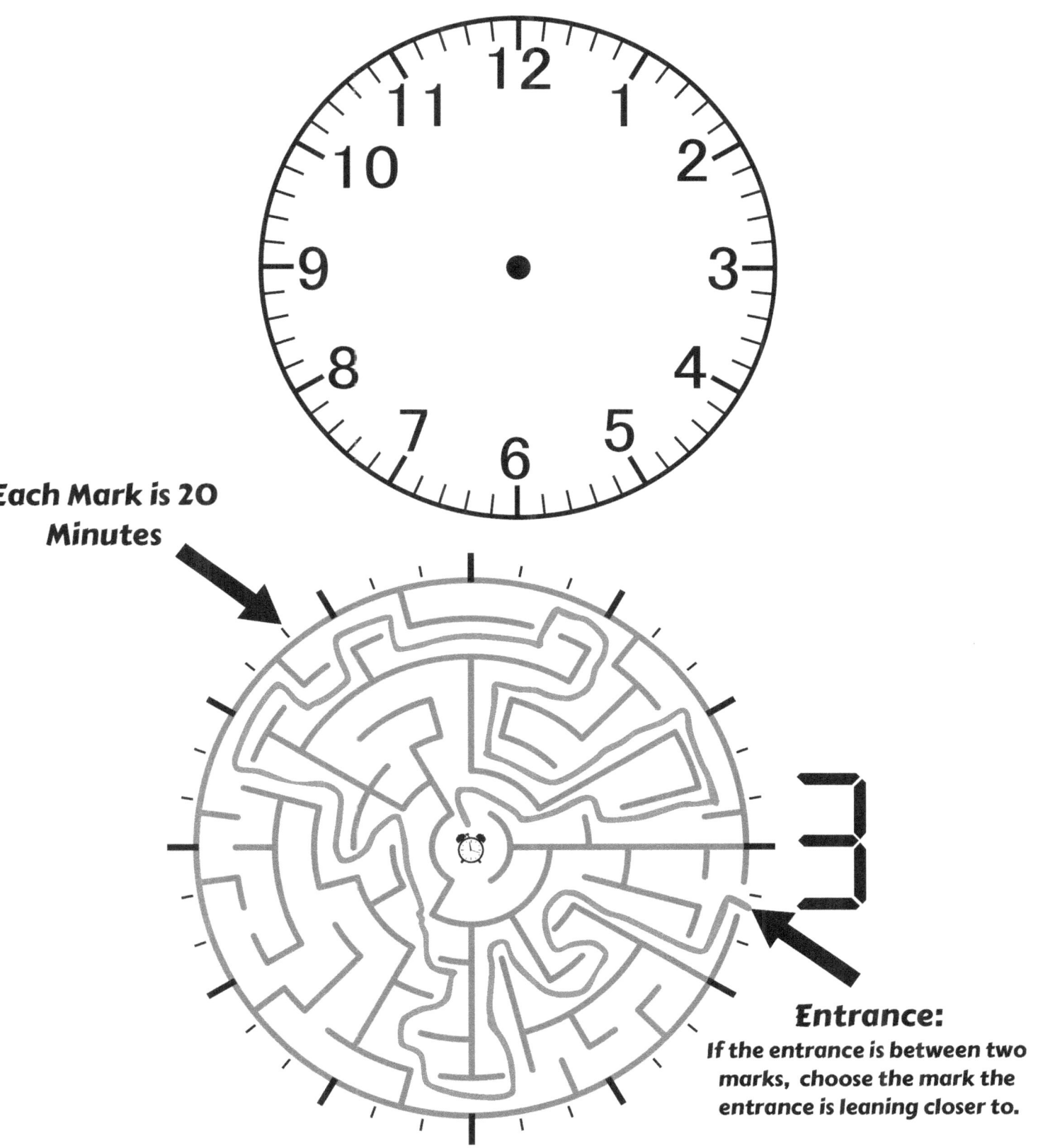

Each Mark is 20 Minutes

Entrance:
If the entrance is between two marks, choose the mark the entrance is leaning closer to.

What Time is the Entrance: 3:20

The solutions for all the times can be found in the back of the book.

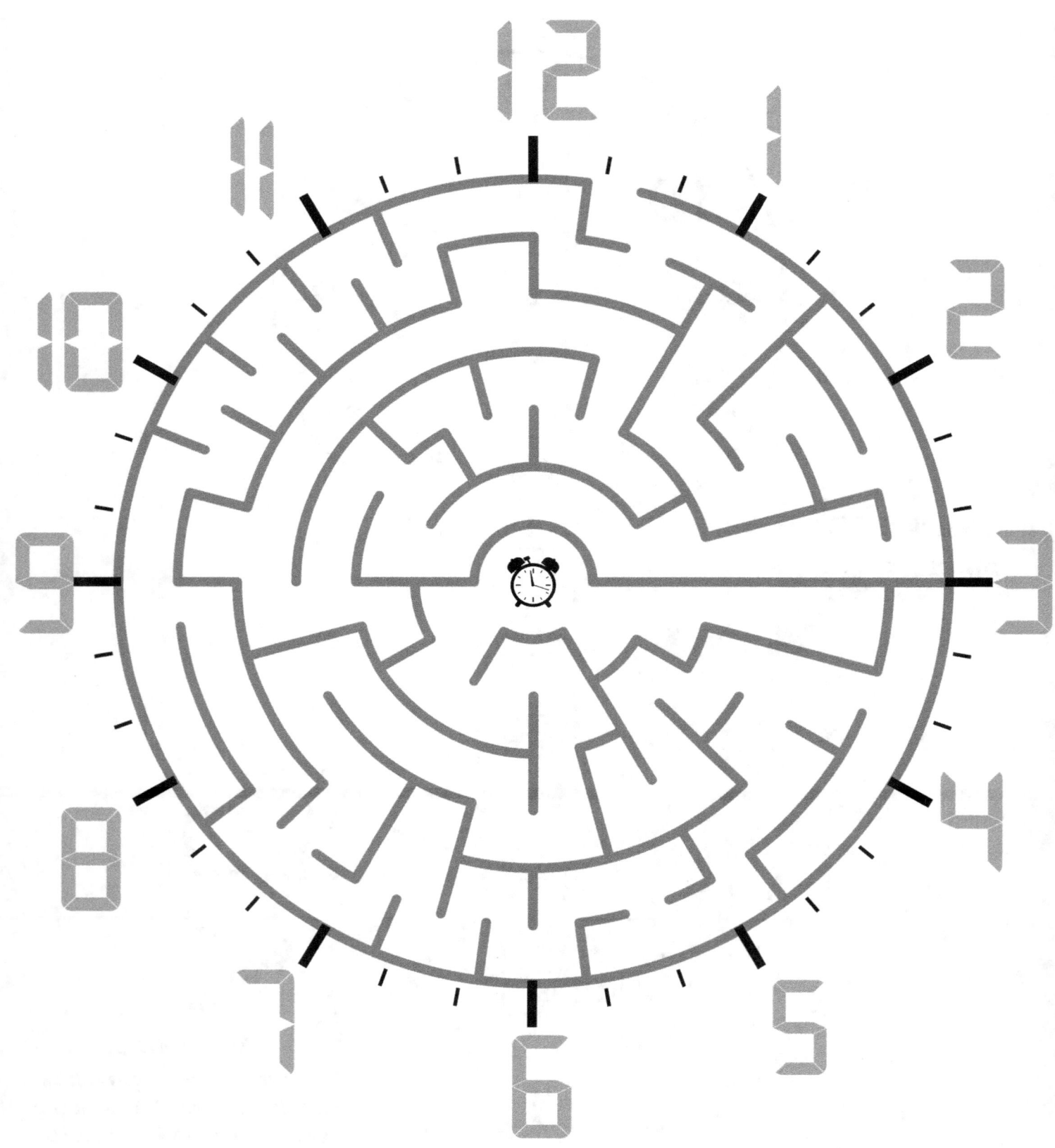

What Time is the Entrance:_____

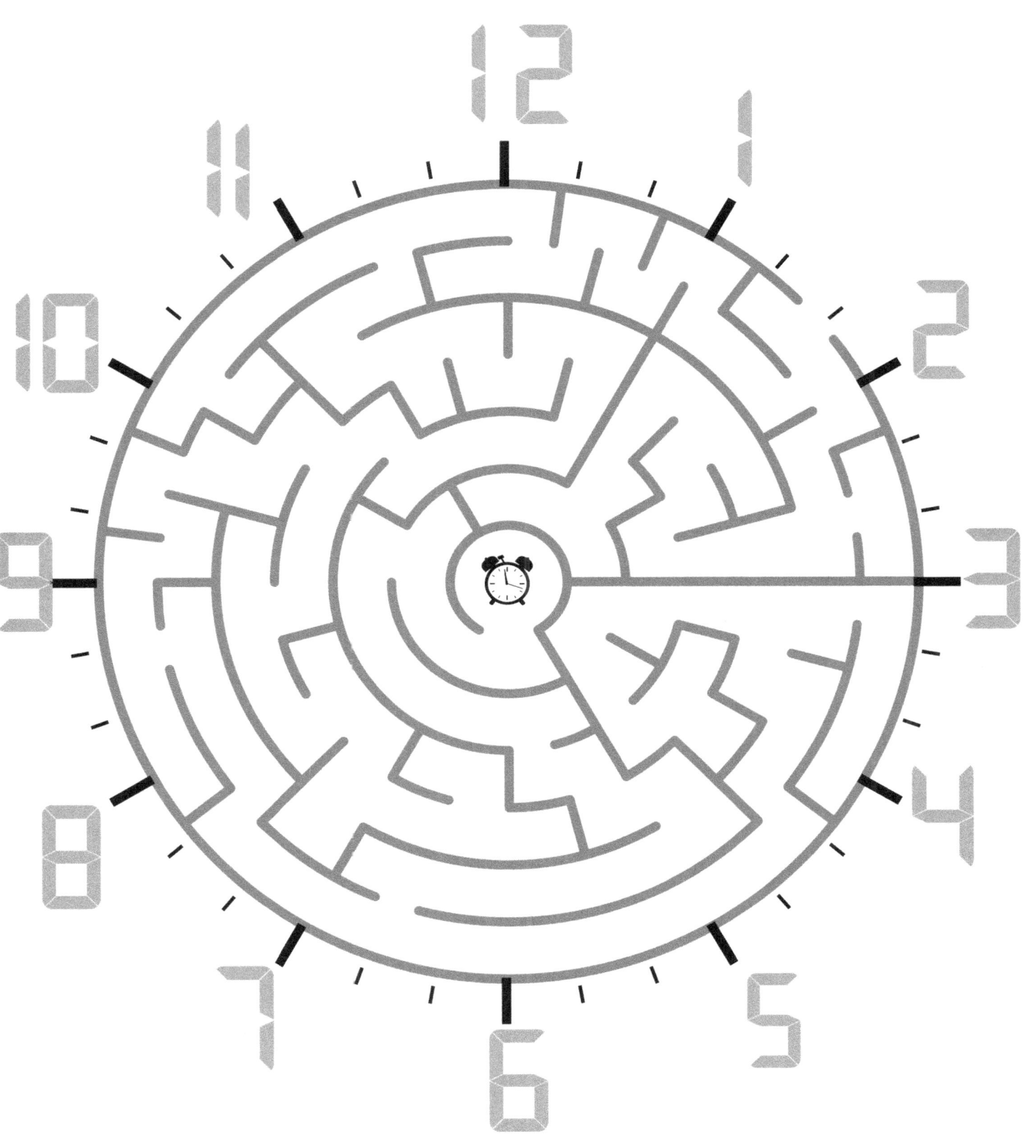

What Time is the Entrance:_____

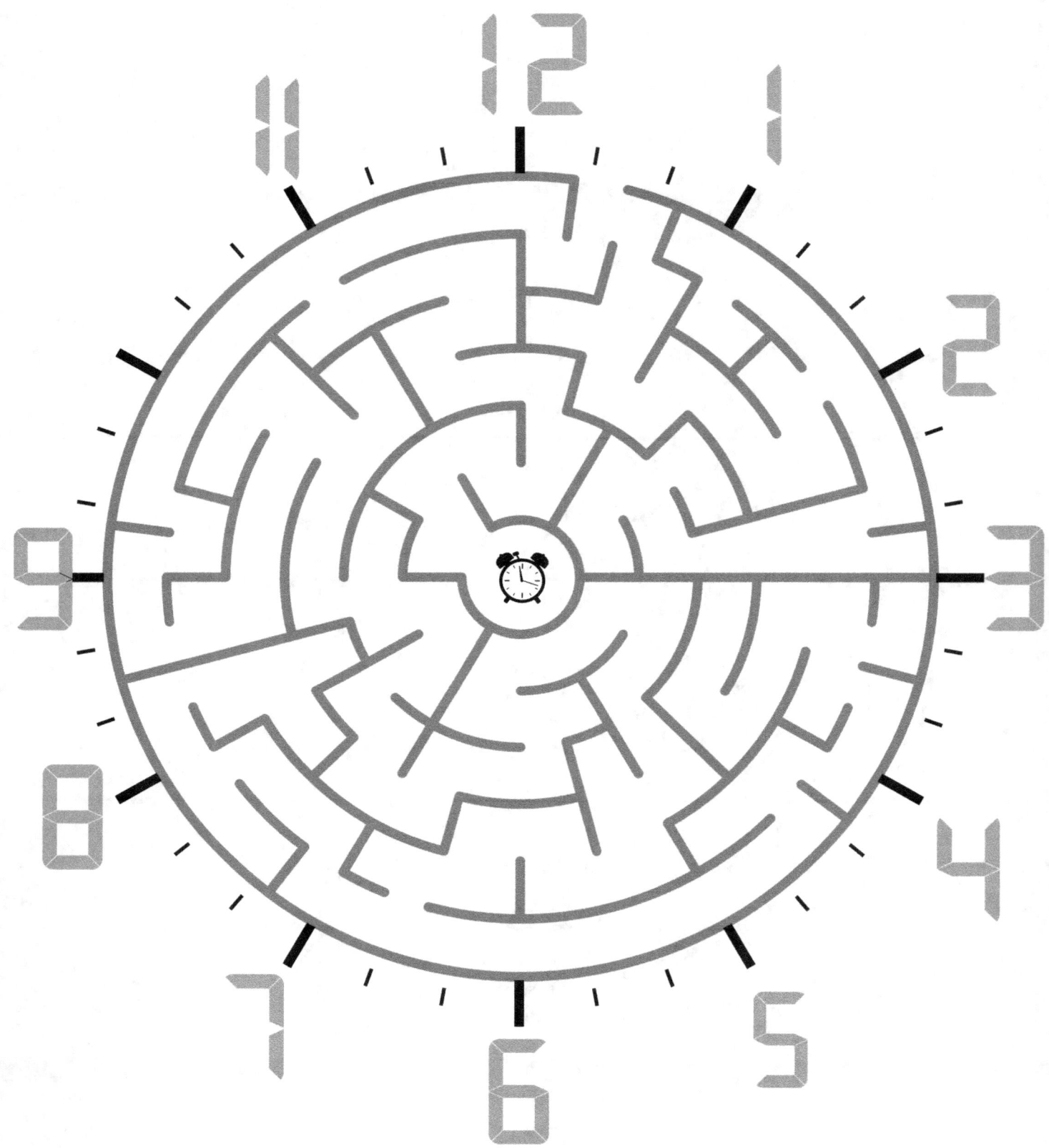

What Time is the Entrance:_____

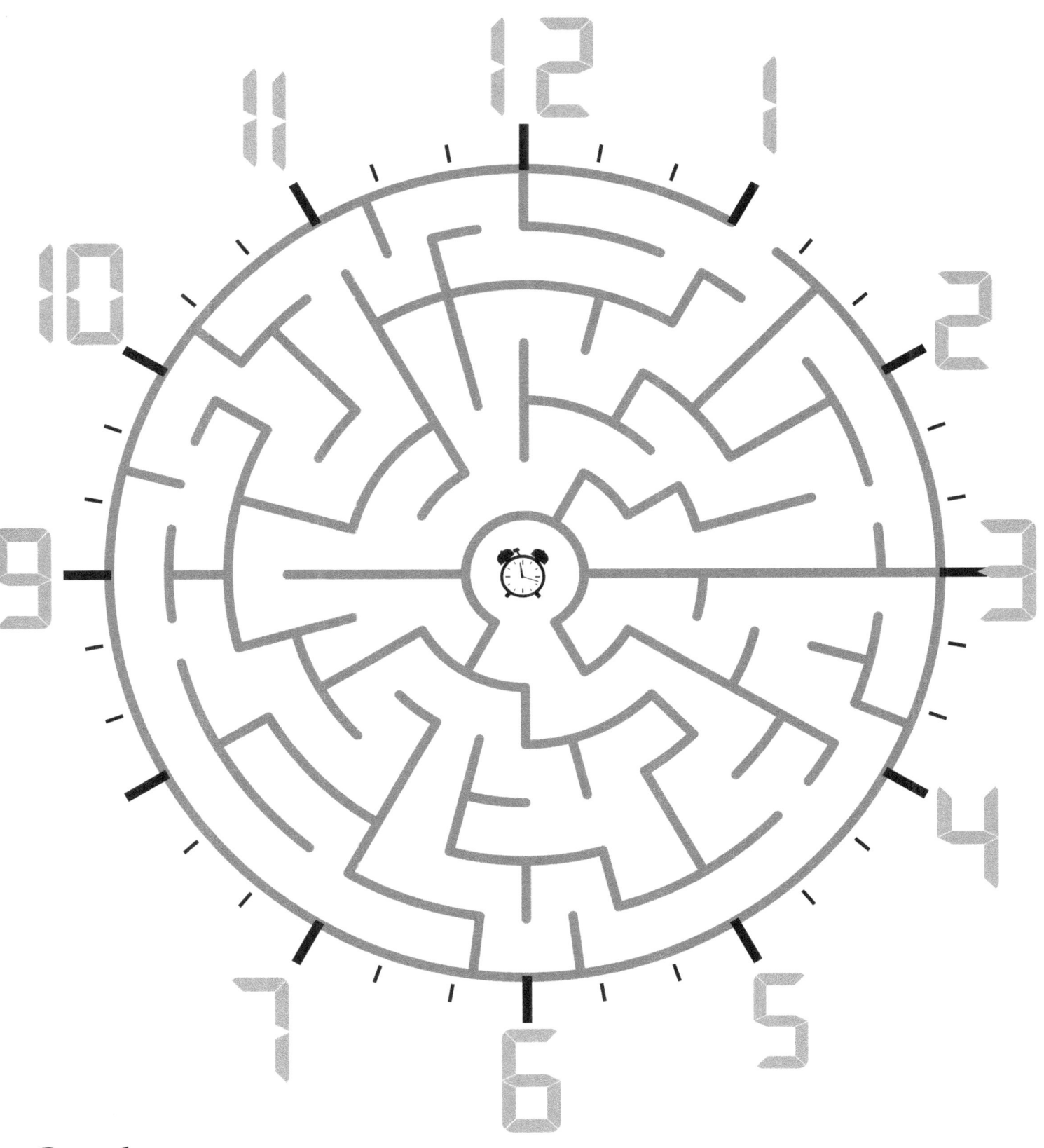

What Time is the Entrance:_____

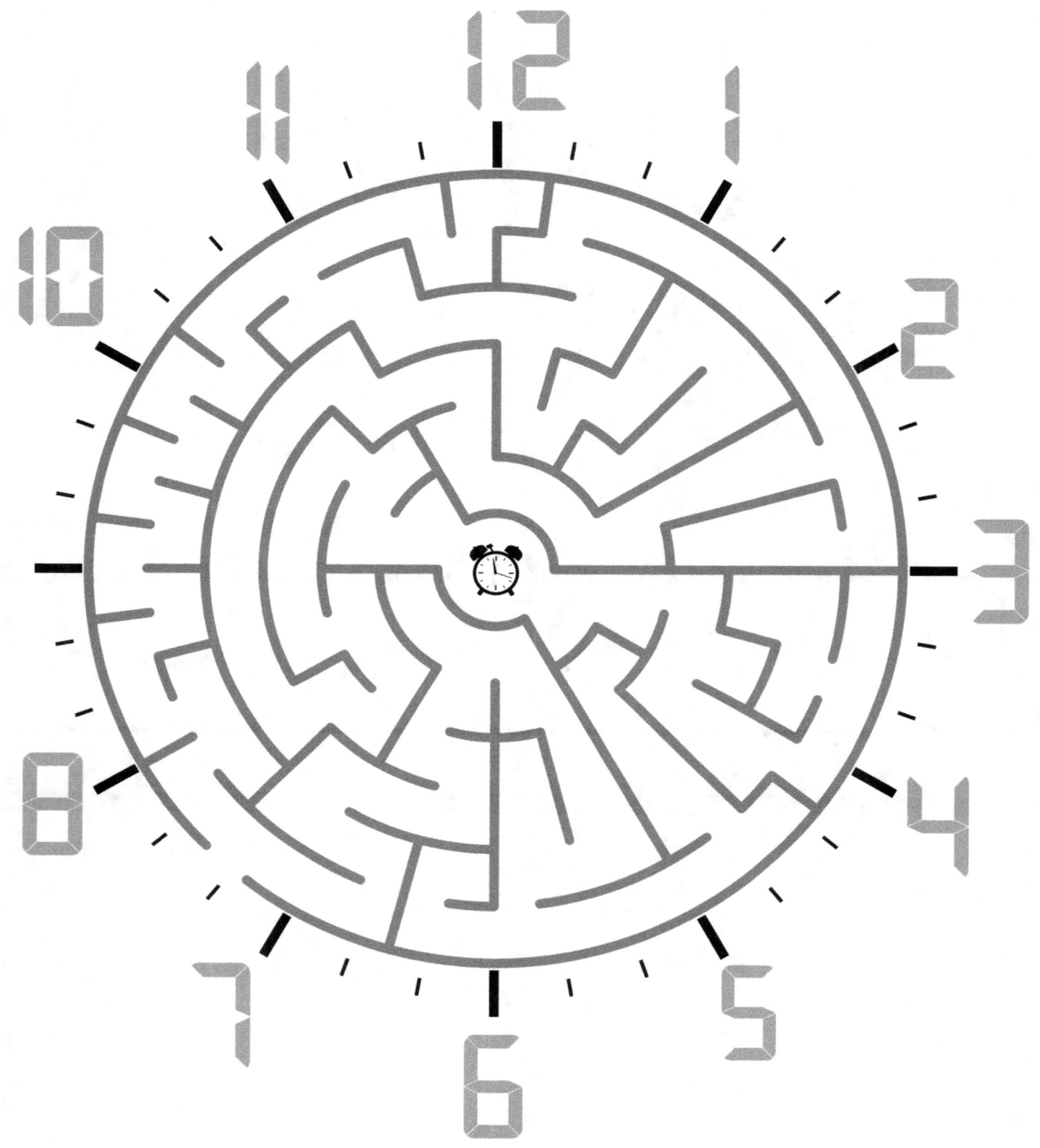

What Time is the Entrance:_____

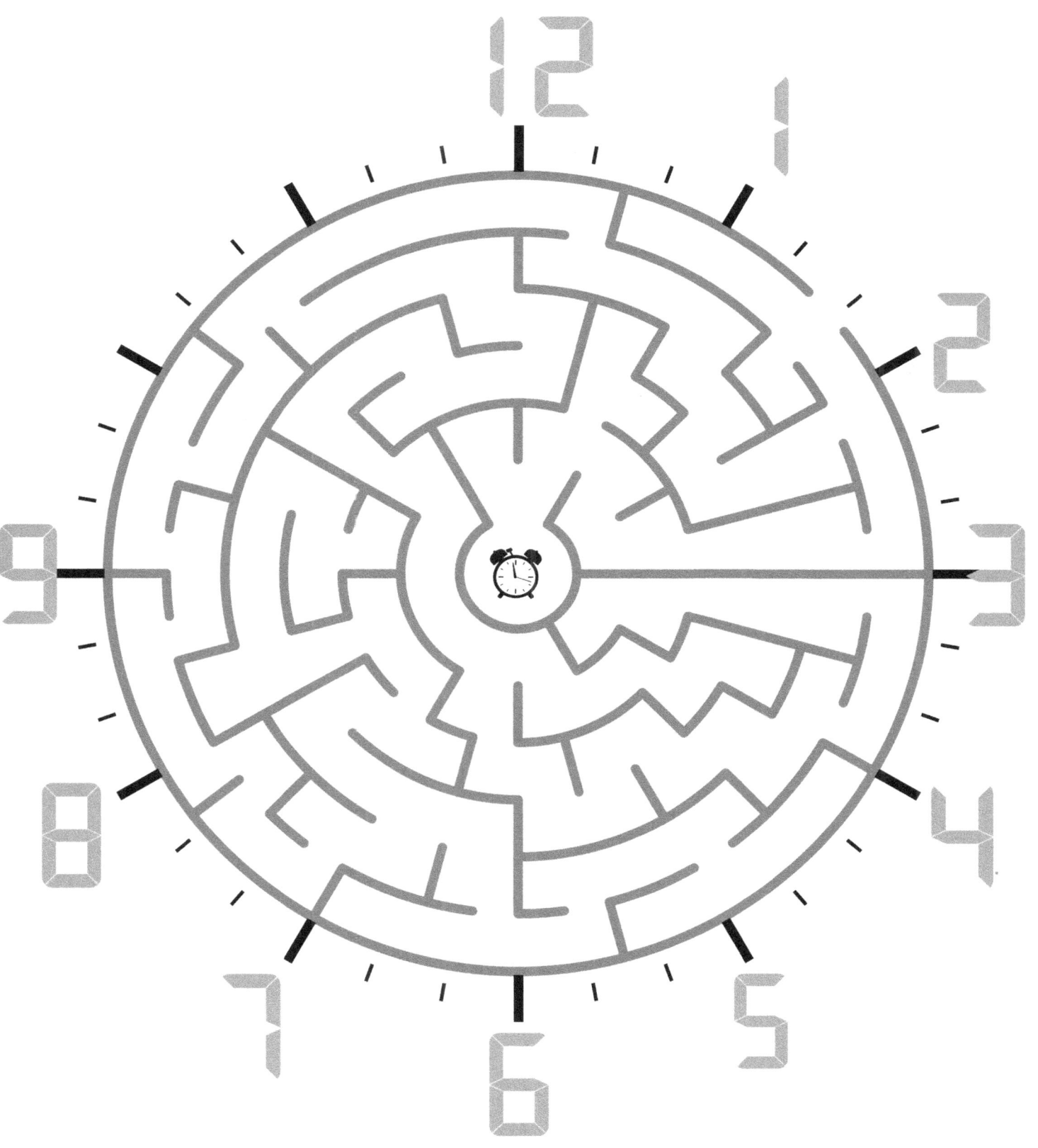

What Time is the Entrance:_____

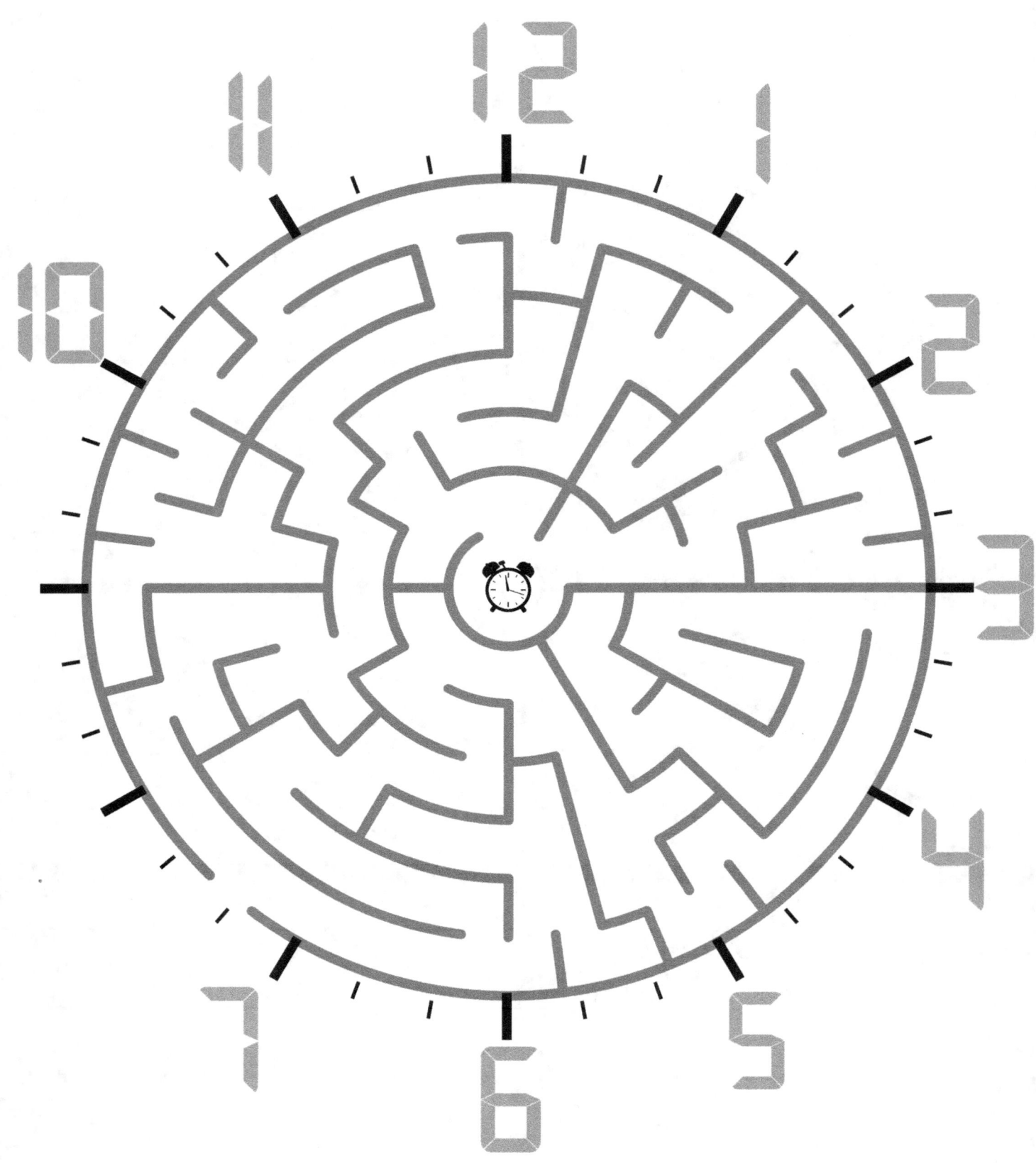

What Time is the Entrance:_____

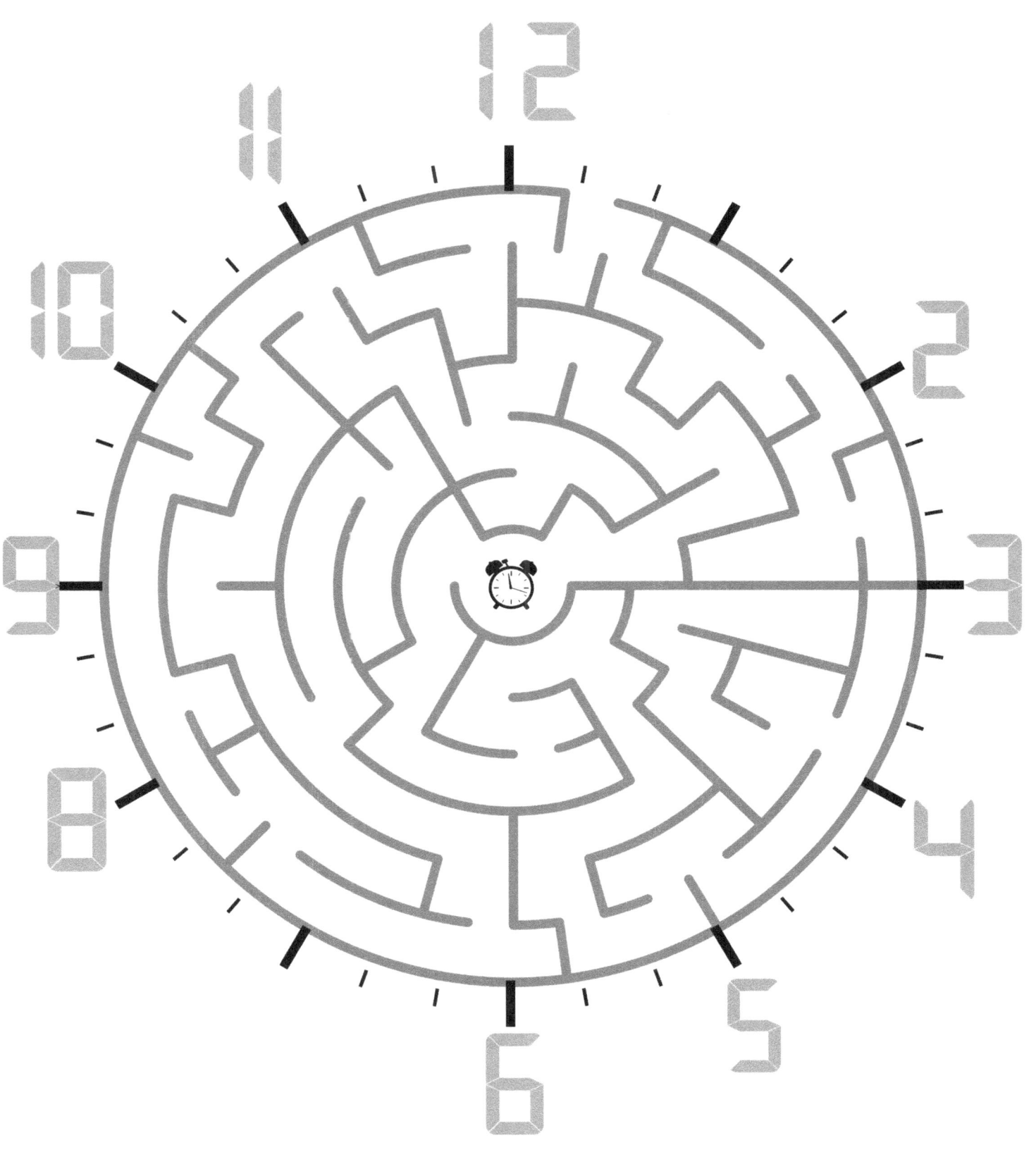

What Time is the Entrance:_____

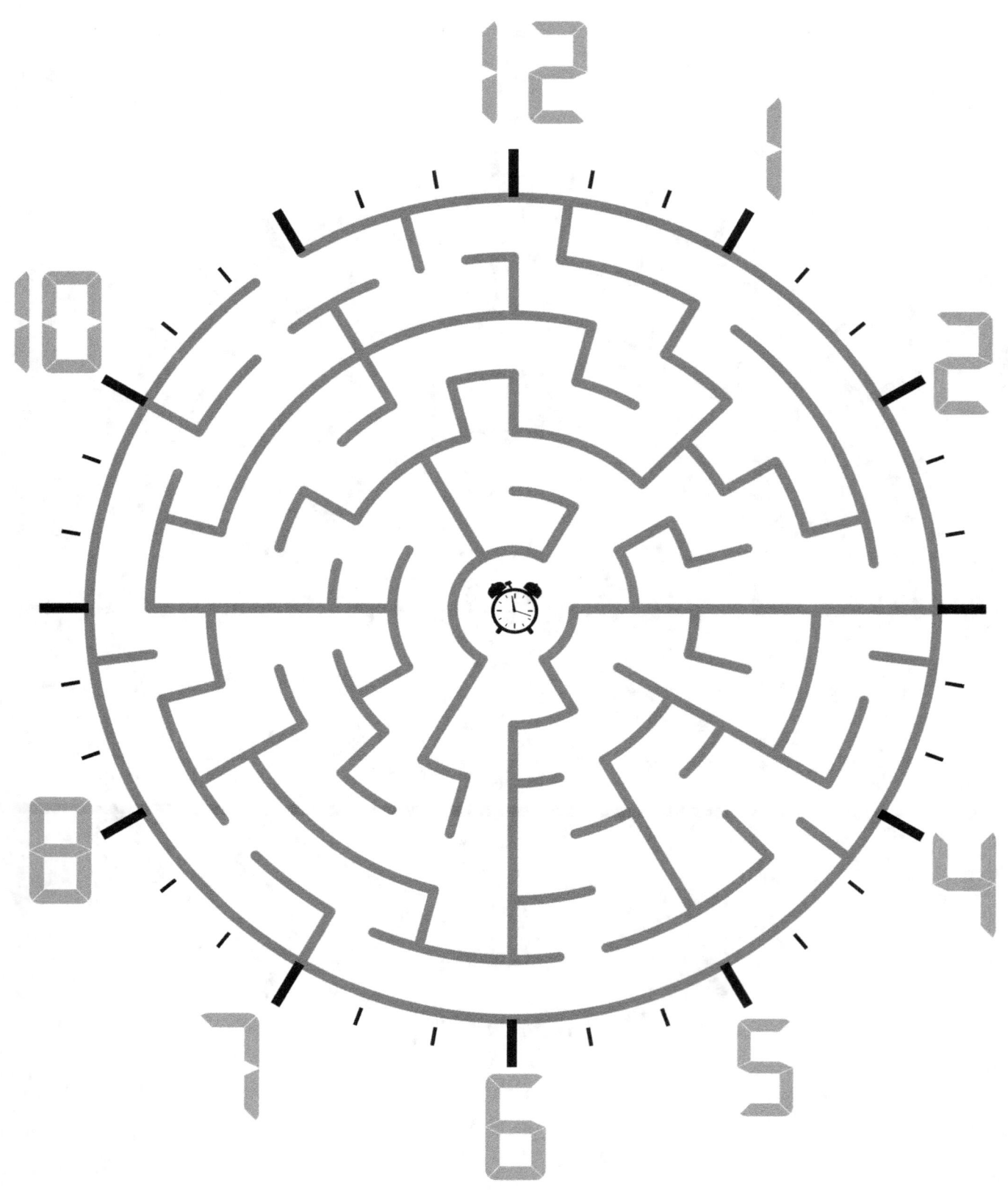

What Time is the Entrance:_____

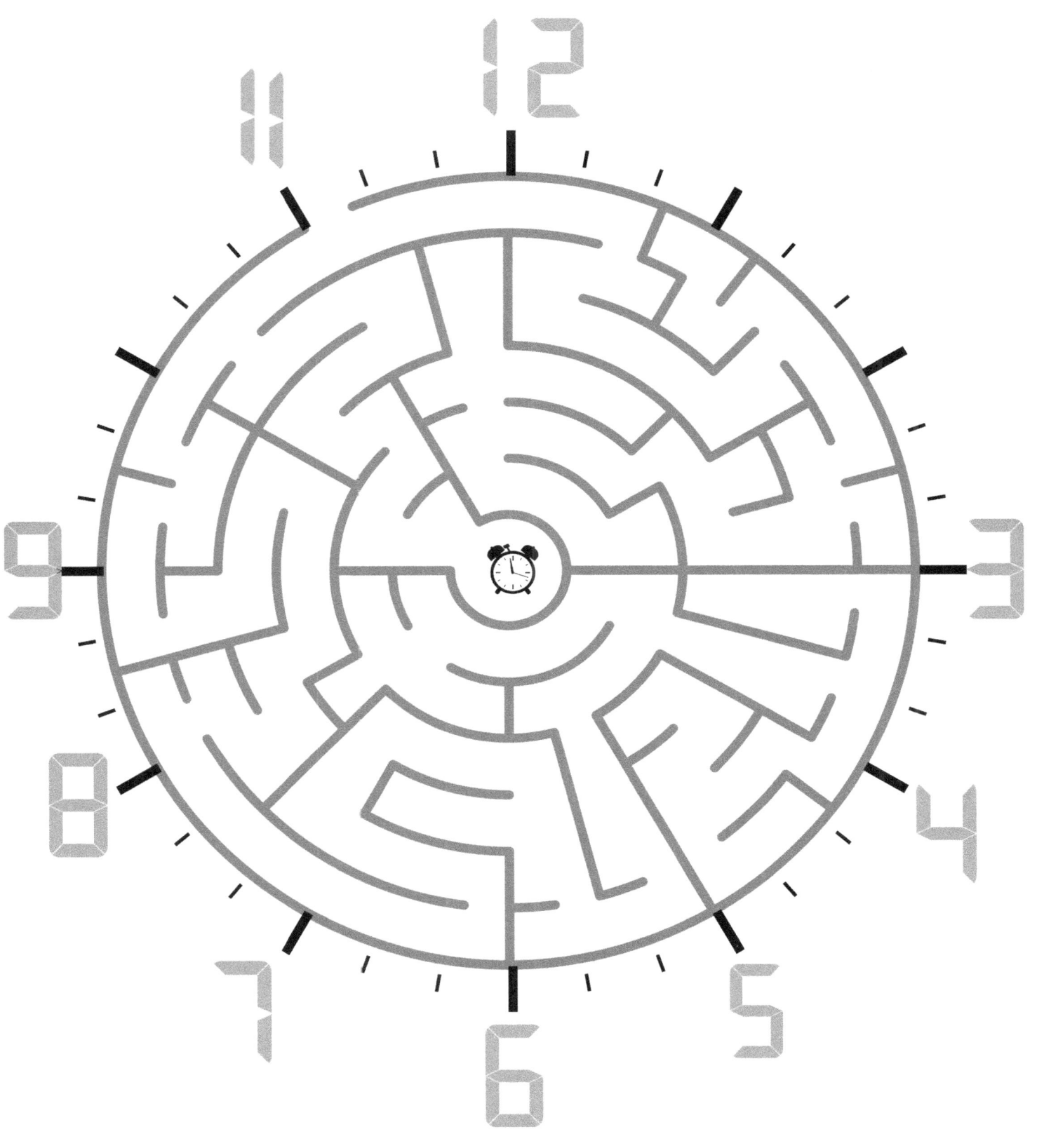

What Time is the Entrance:_____

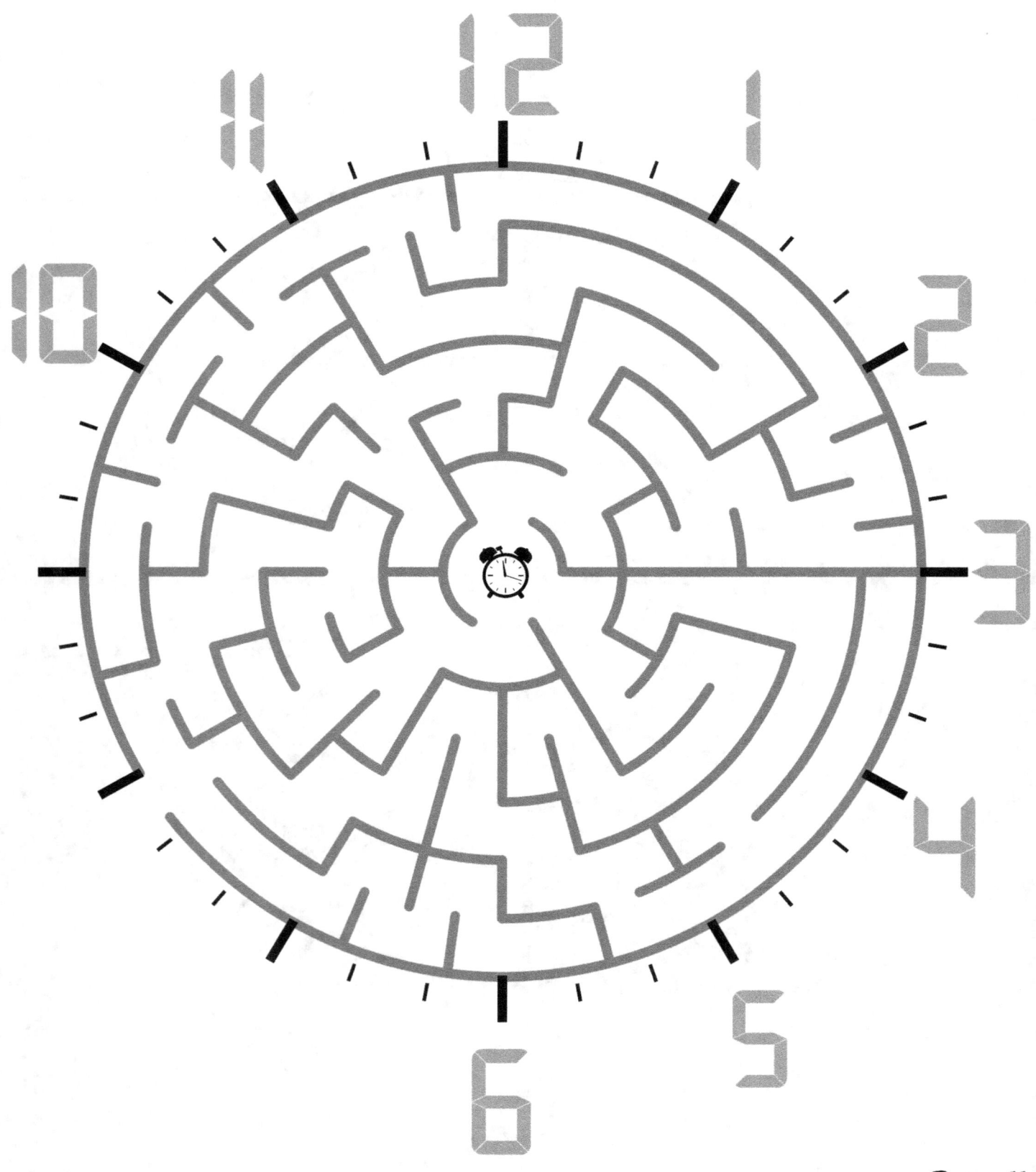

What Time is the Entrance:_____

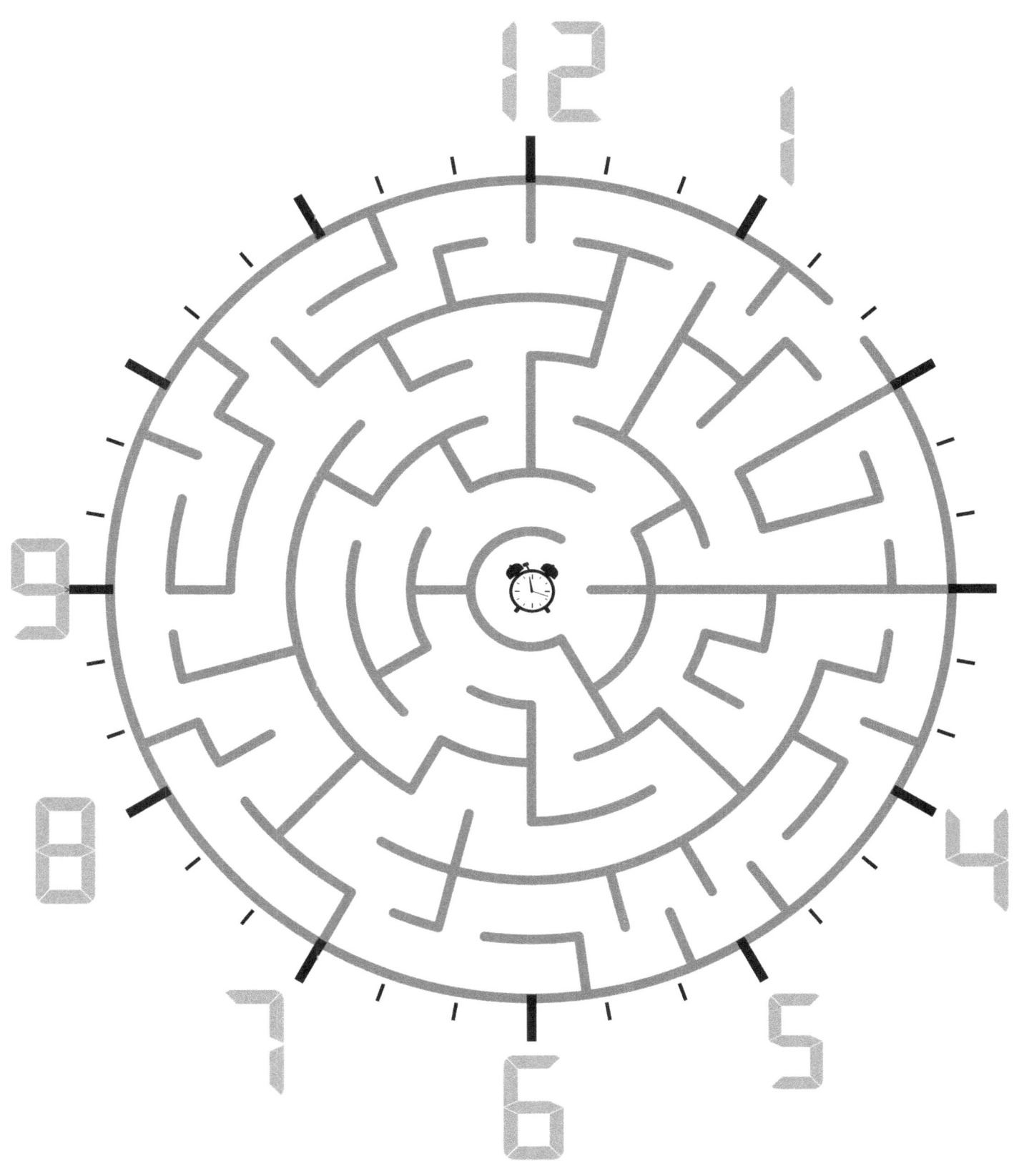

What Time is the Entrance:_____

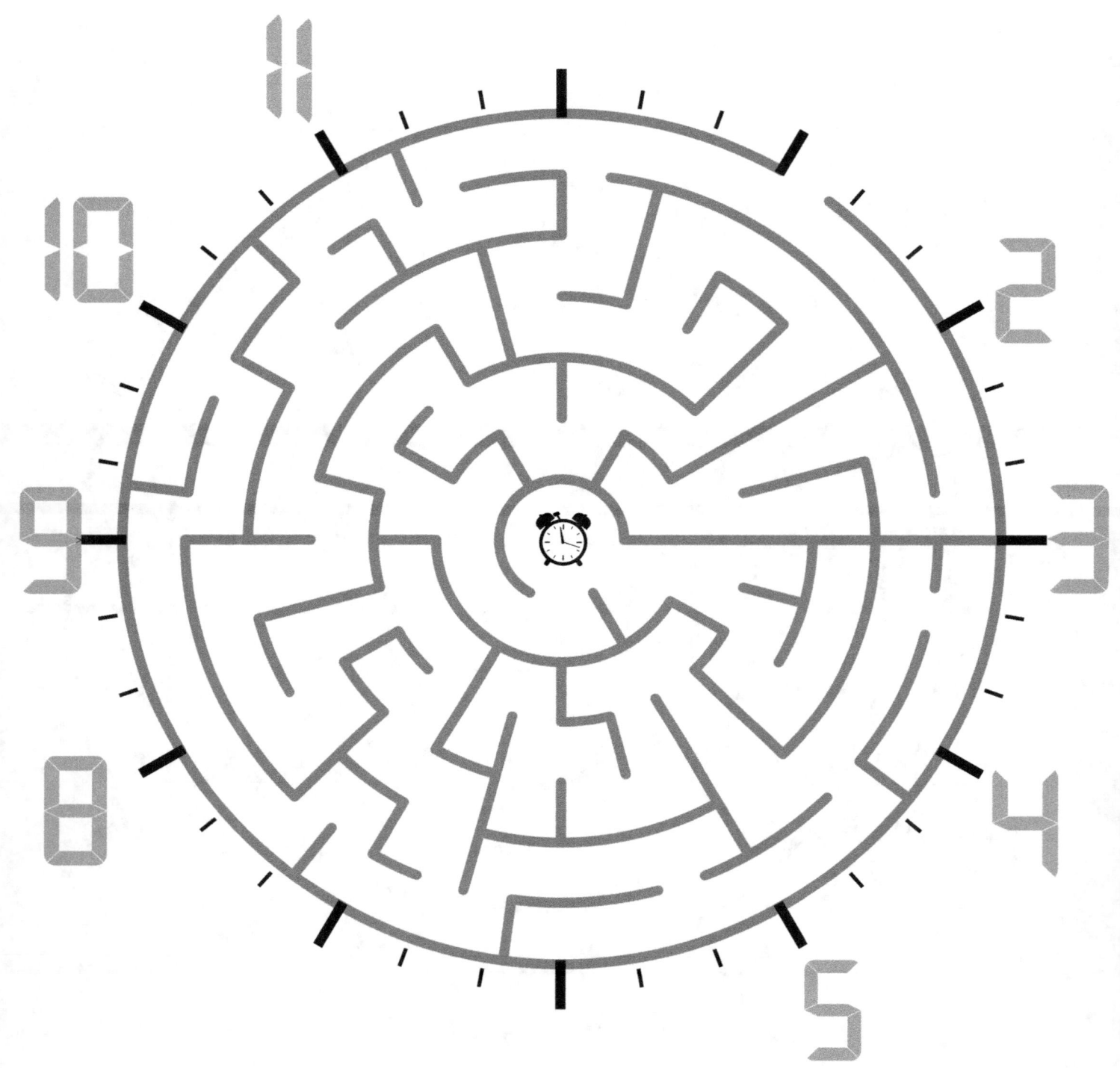

What Time is the Entrance:_____

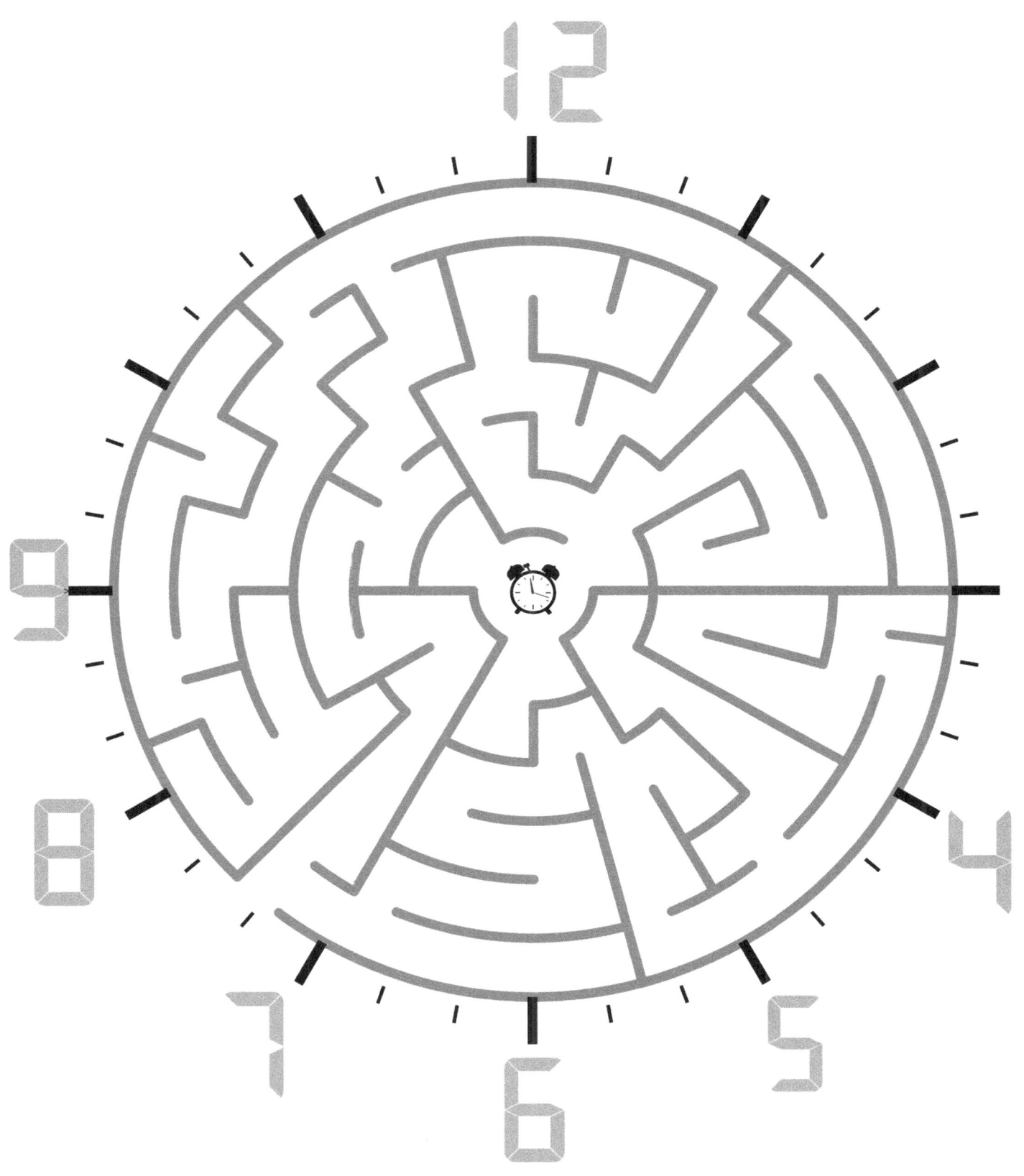

What Time is the Entrance:_____

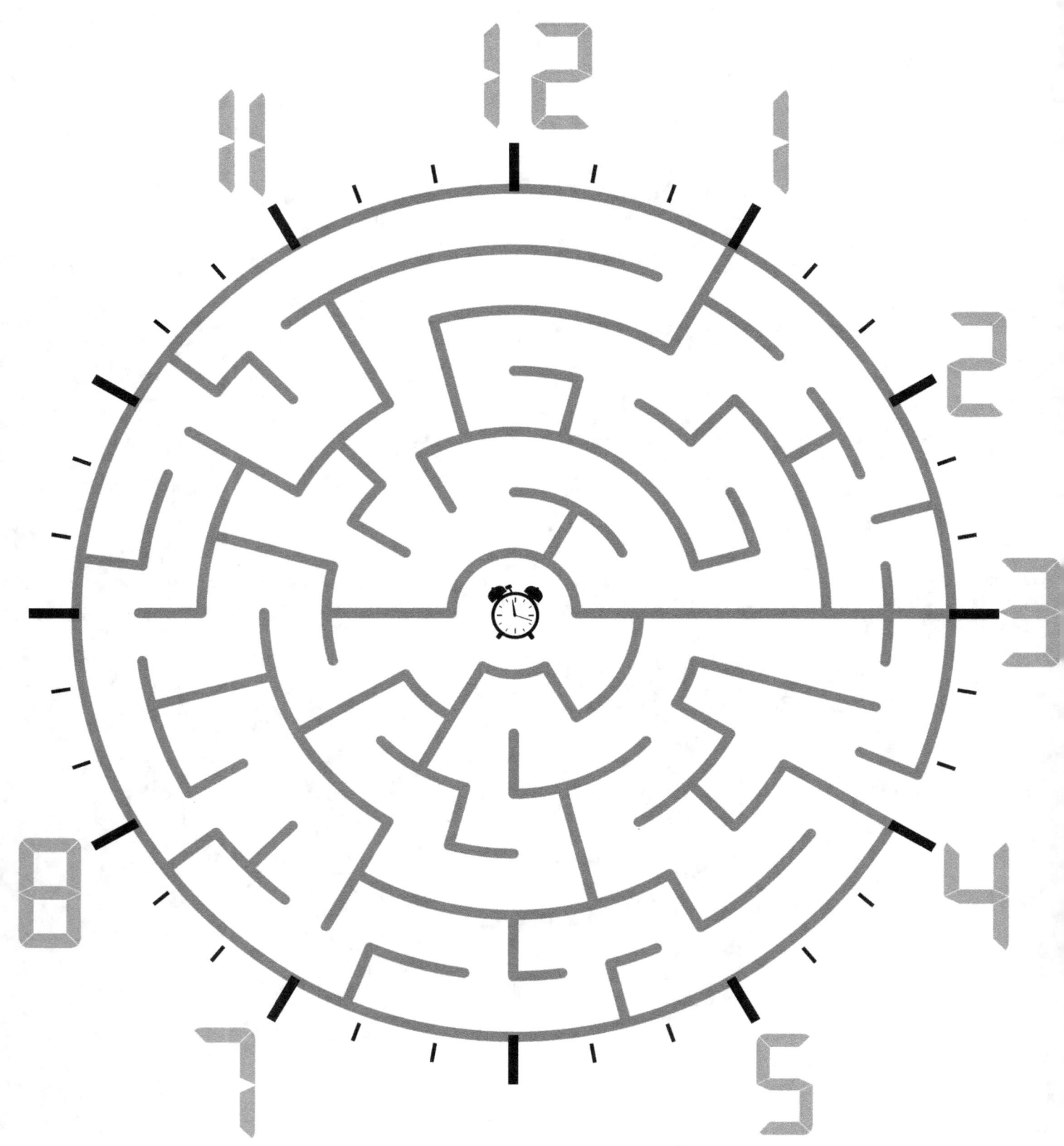

What Time is the Entrance:_____

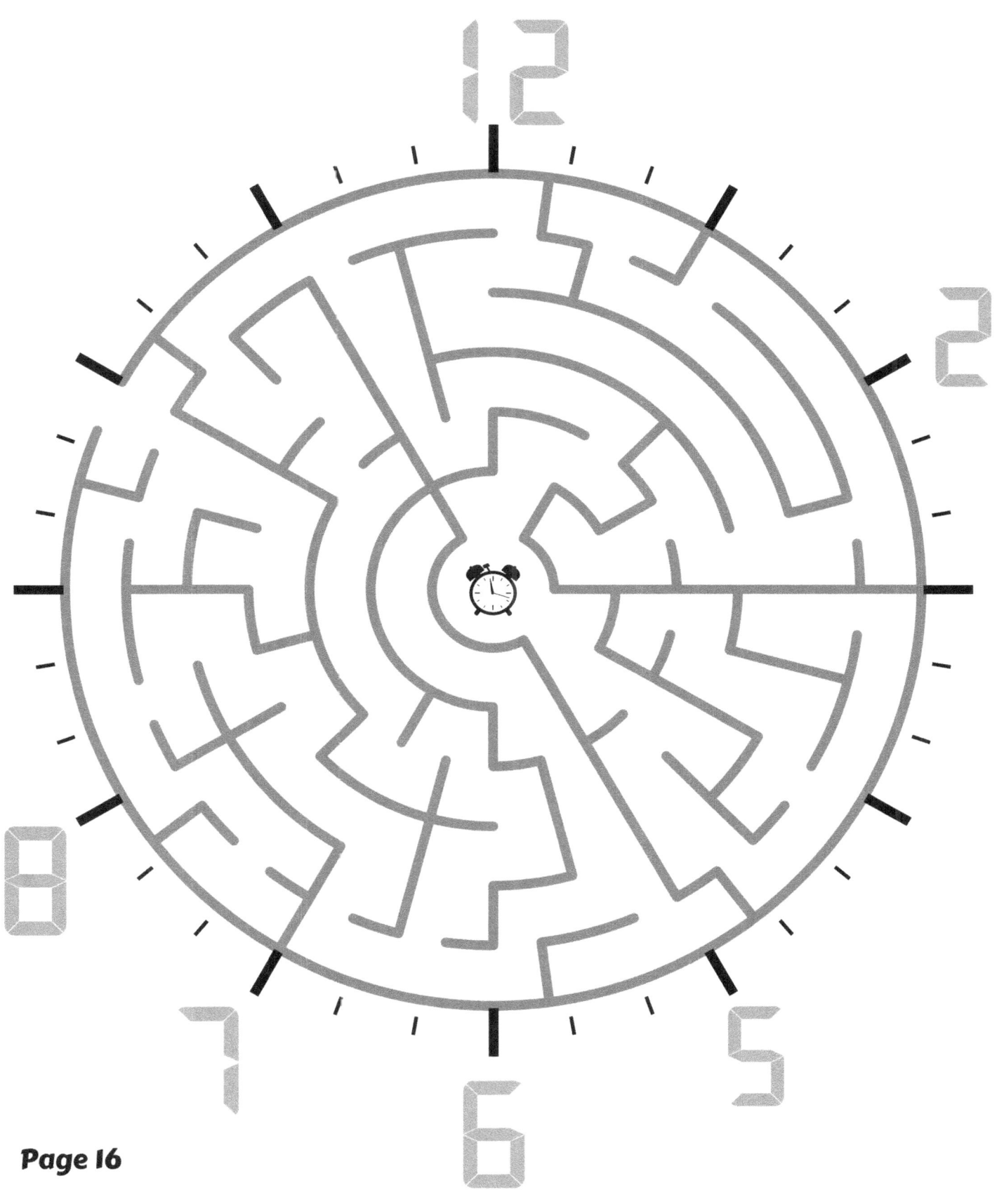

What Time is the Entrance:_____

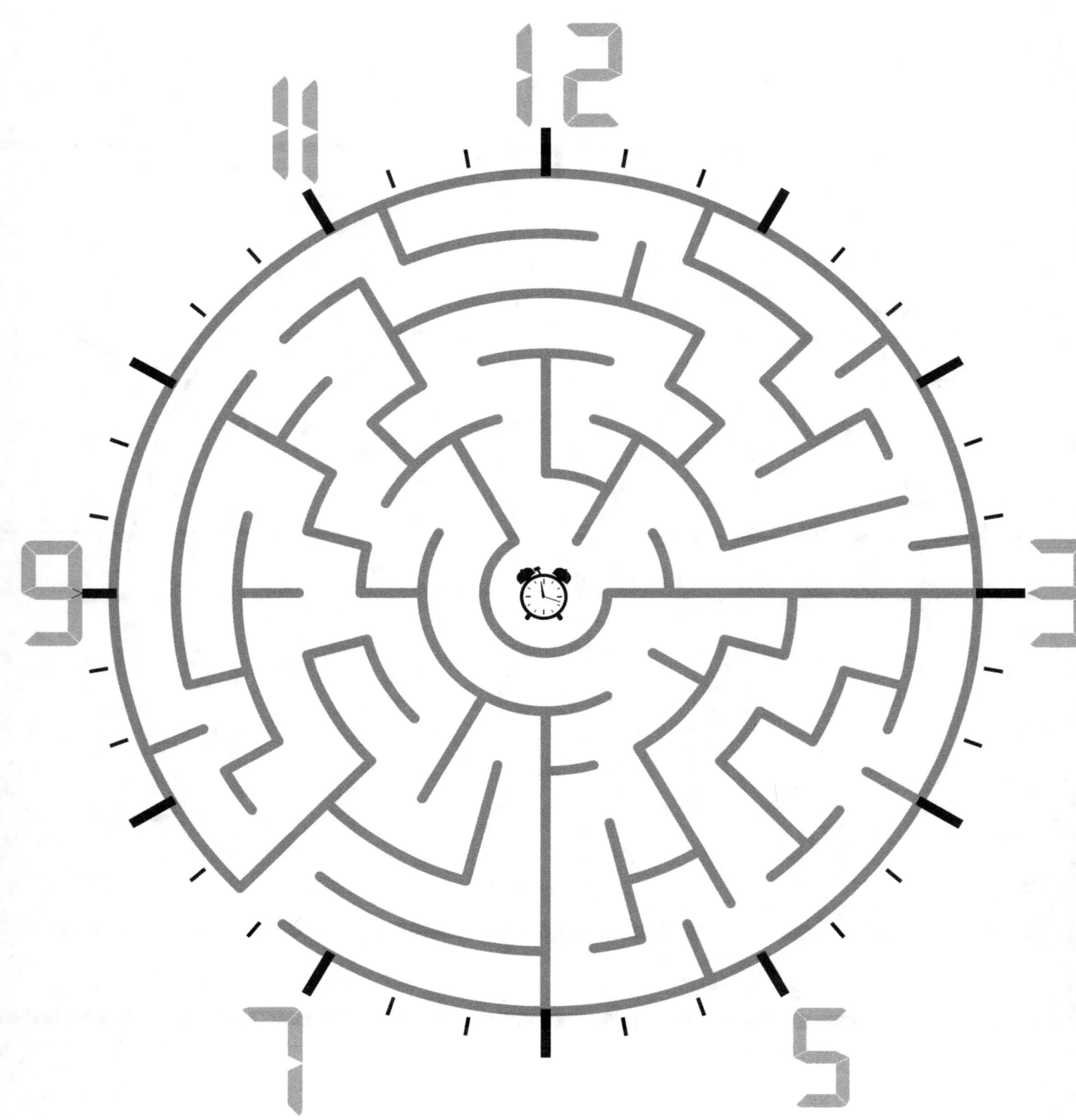

What Time is the Entrance:_____

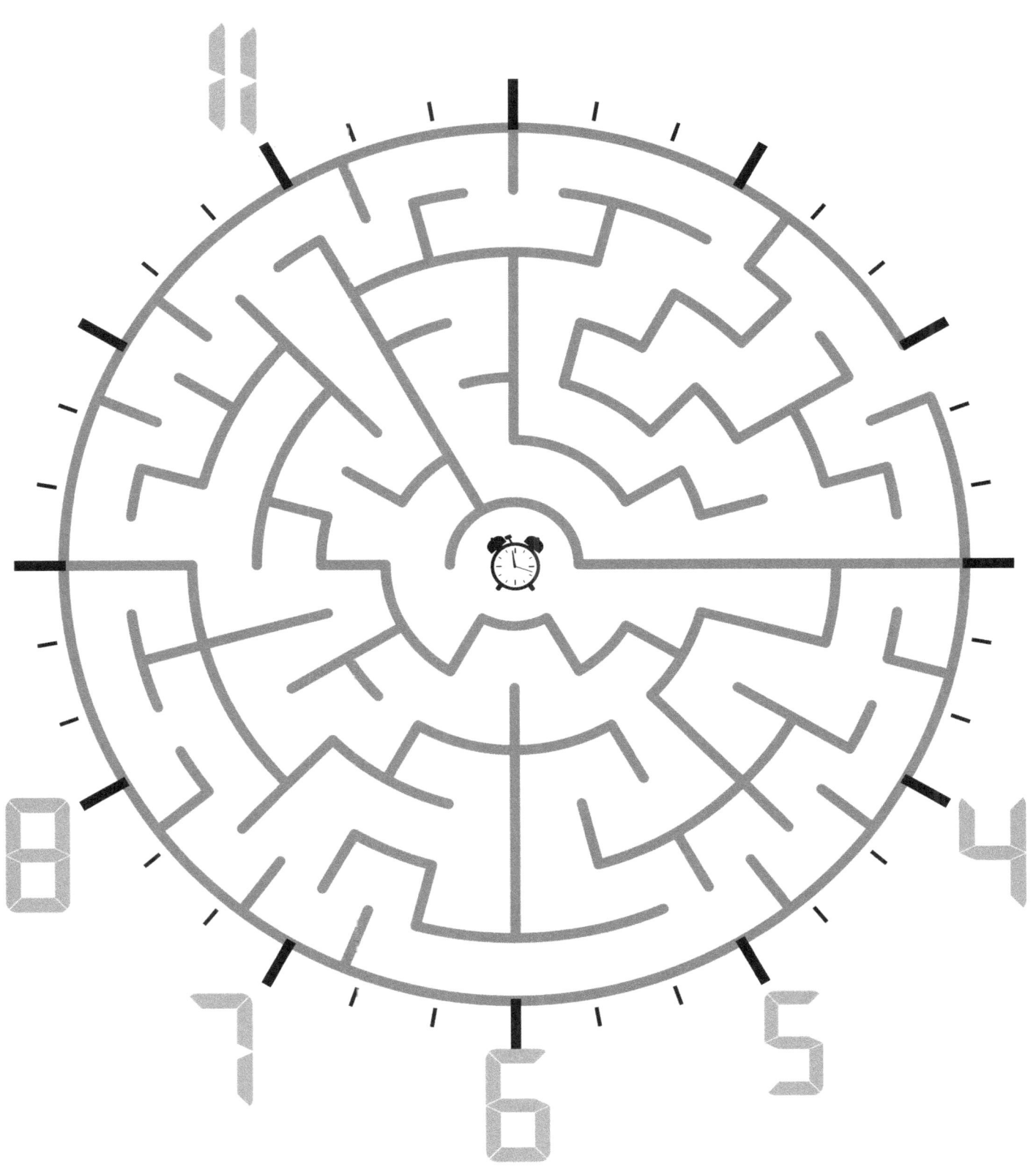

What Time is the Entrance:_____

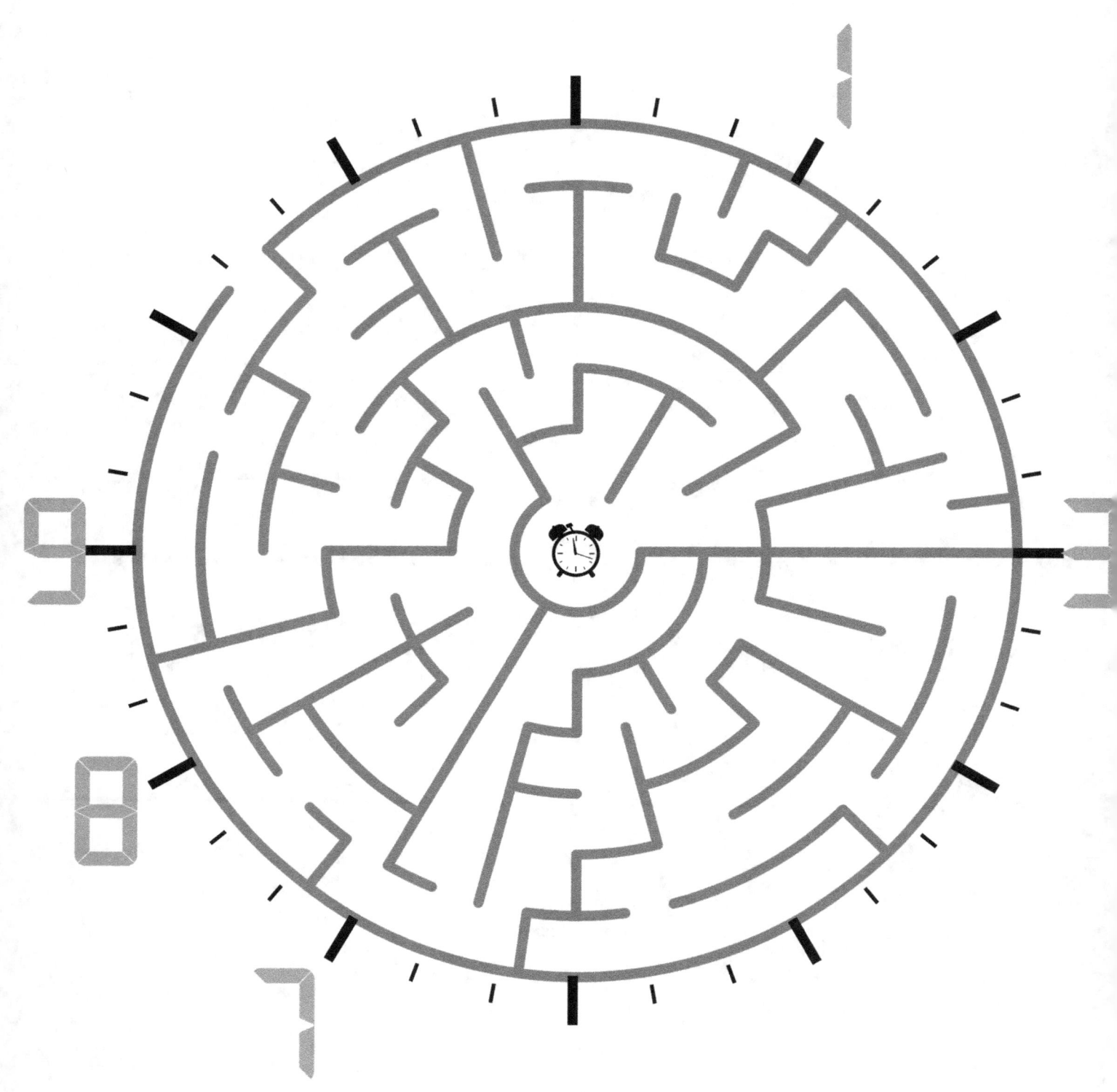

What Time is the Entrance:_____

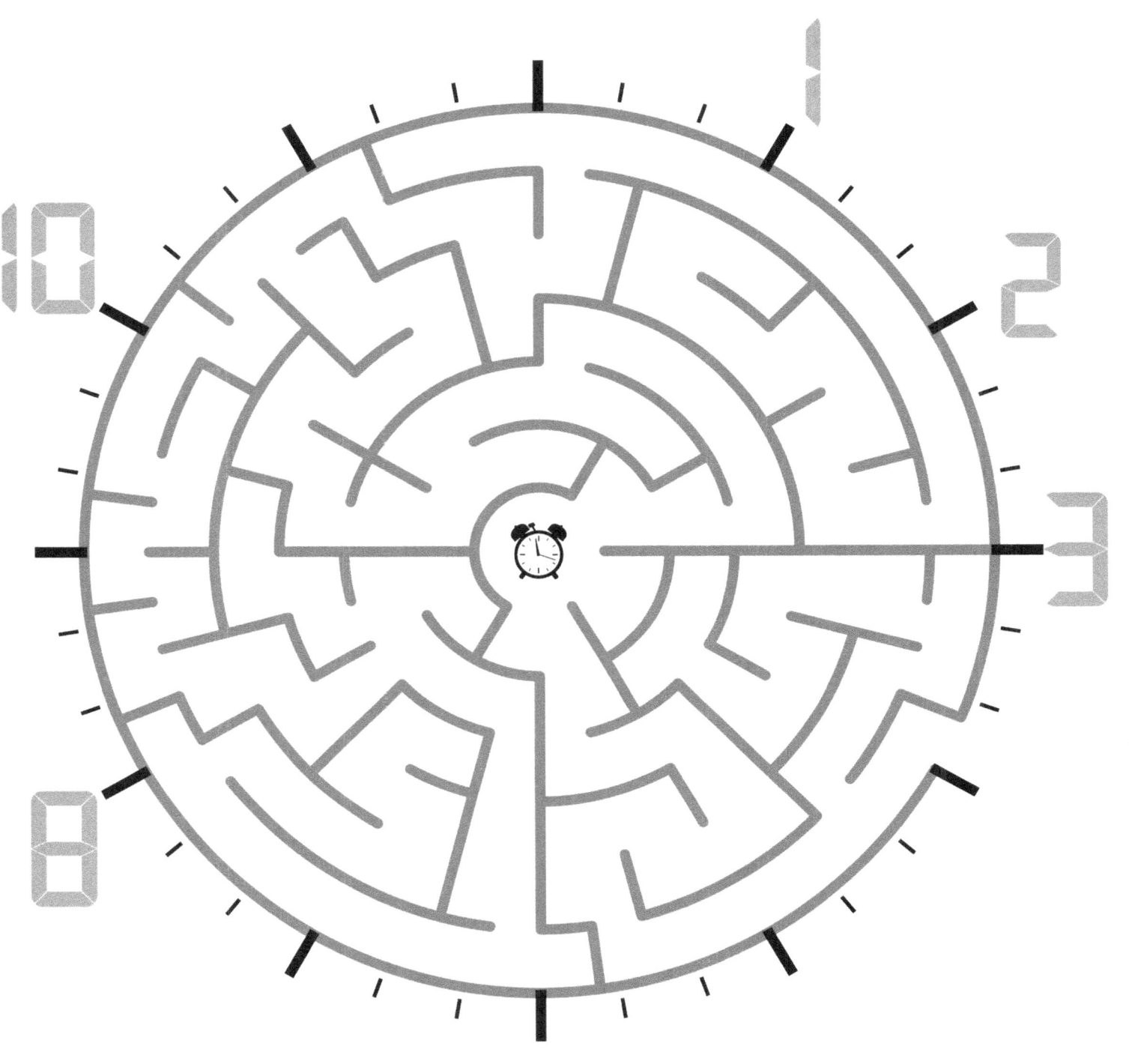

What Time is the Entrance:_____

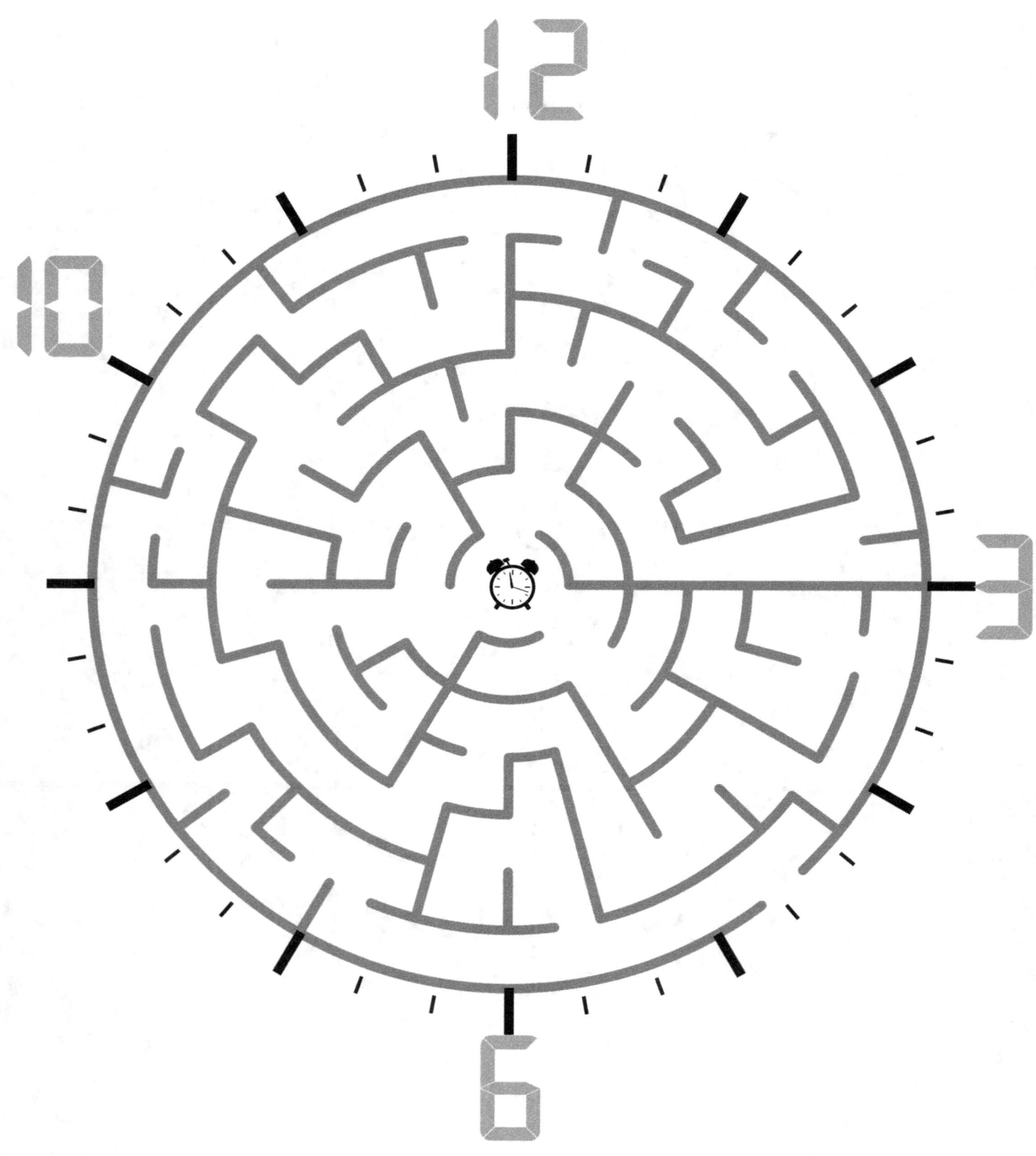

What Time is the Entrance:_____

What Time is the Entrance:_____

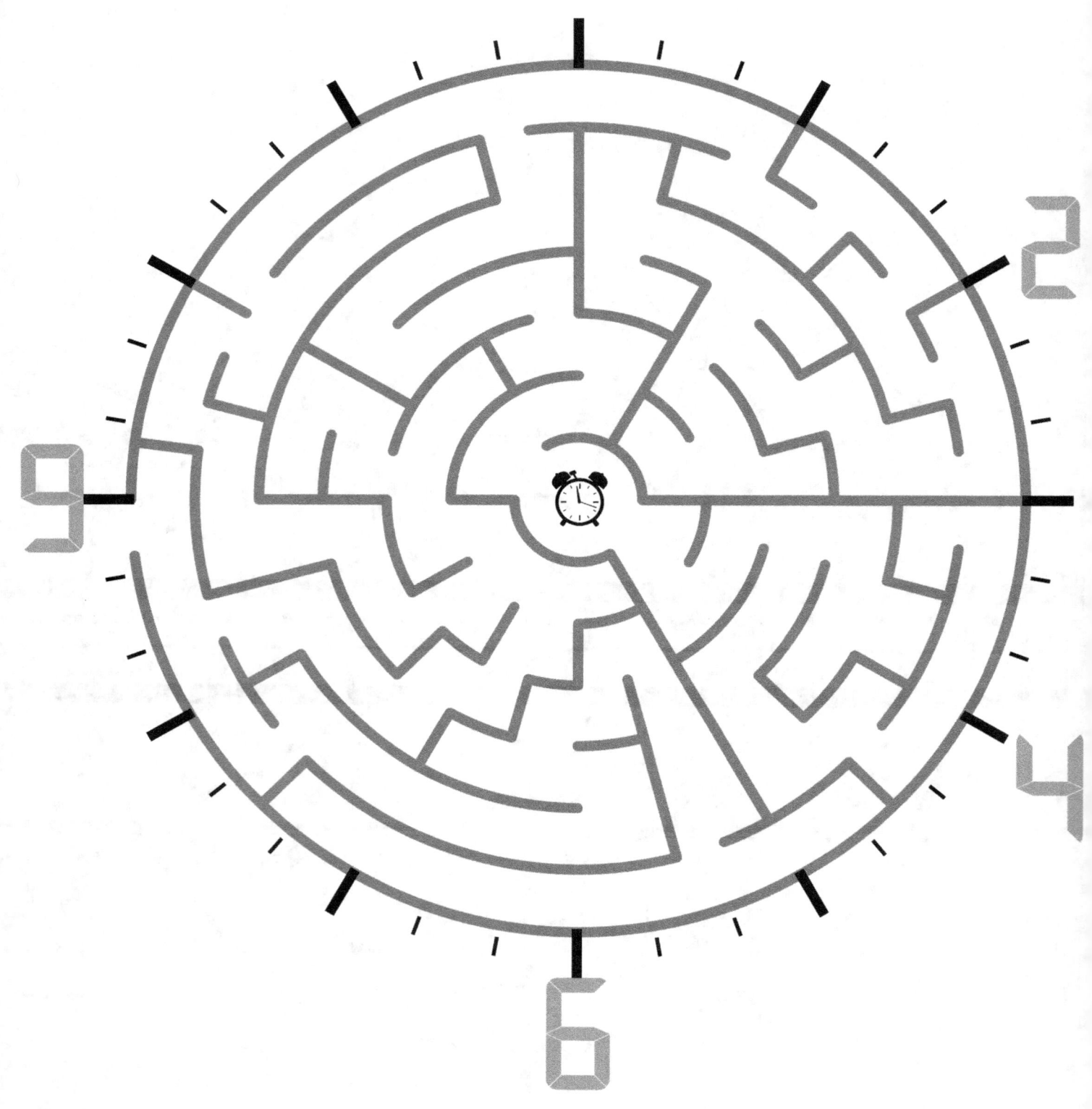

What Time is the Entrance:_____

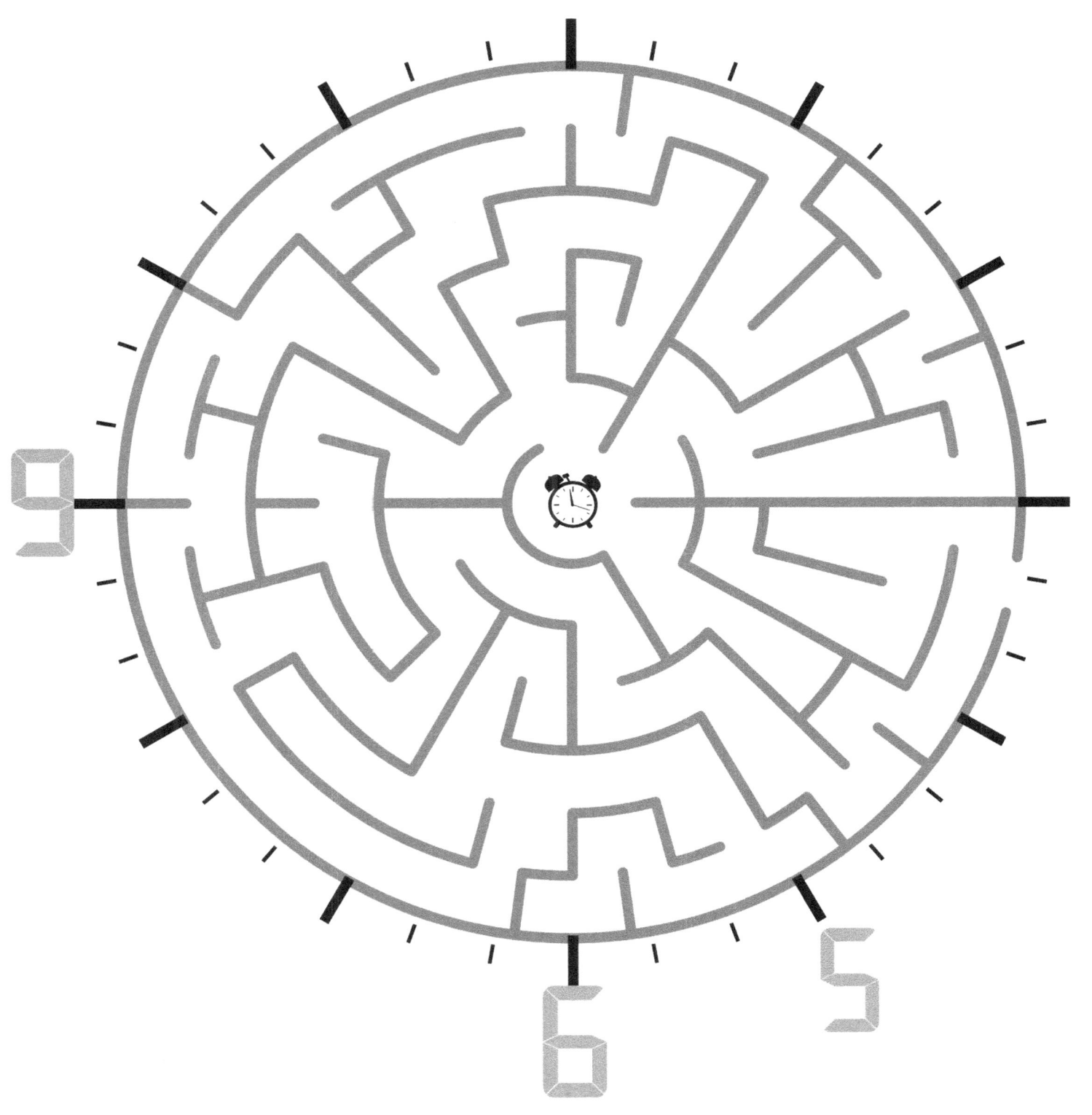

What Time is the Entrance:_____

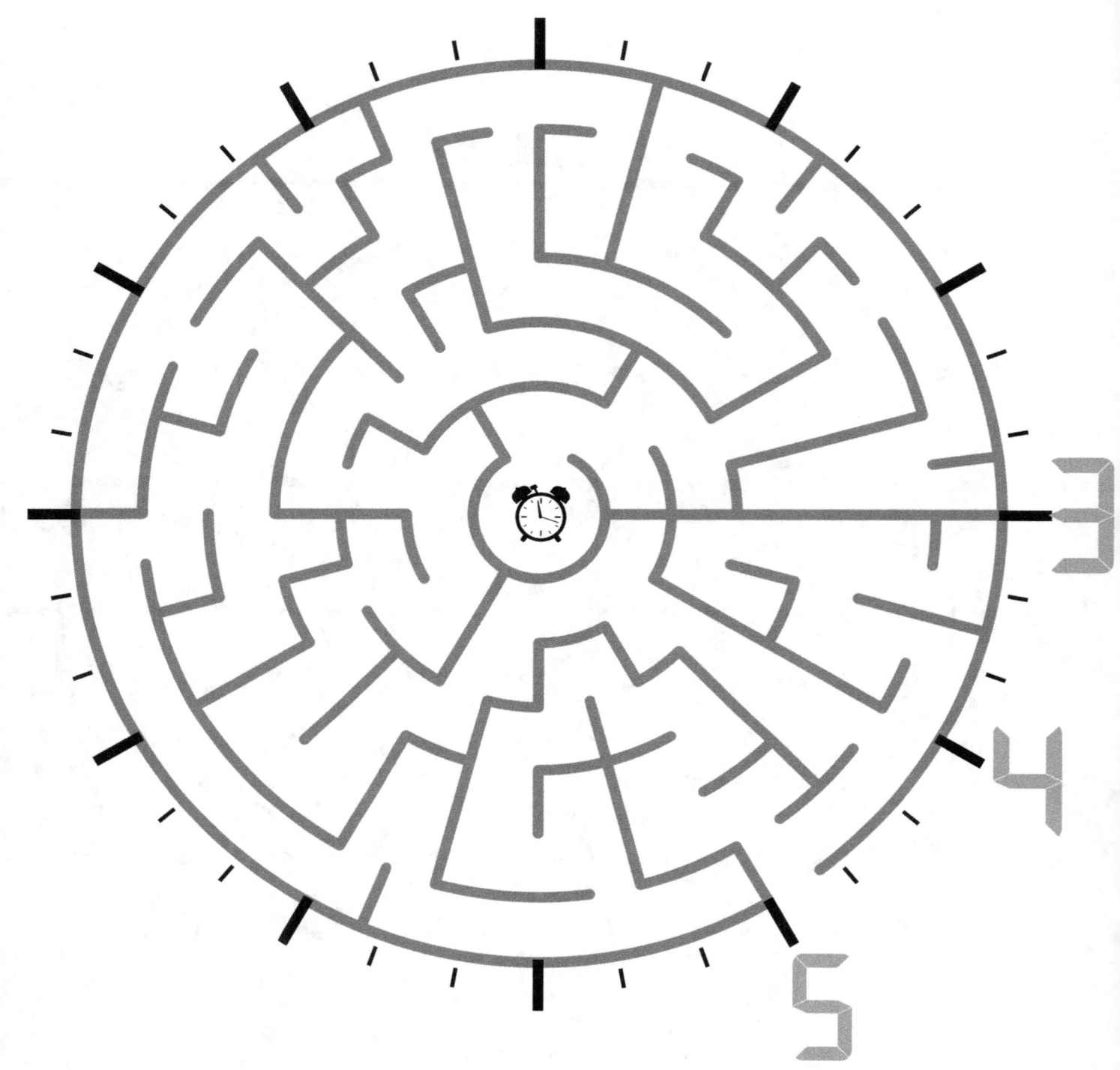

What Time is the Entrance:_____

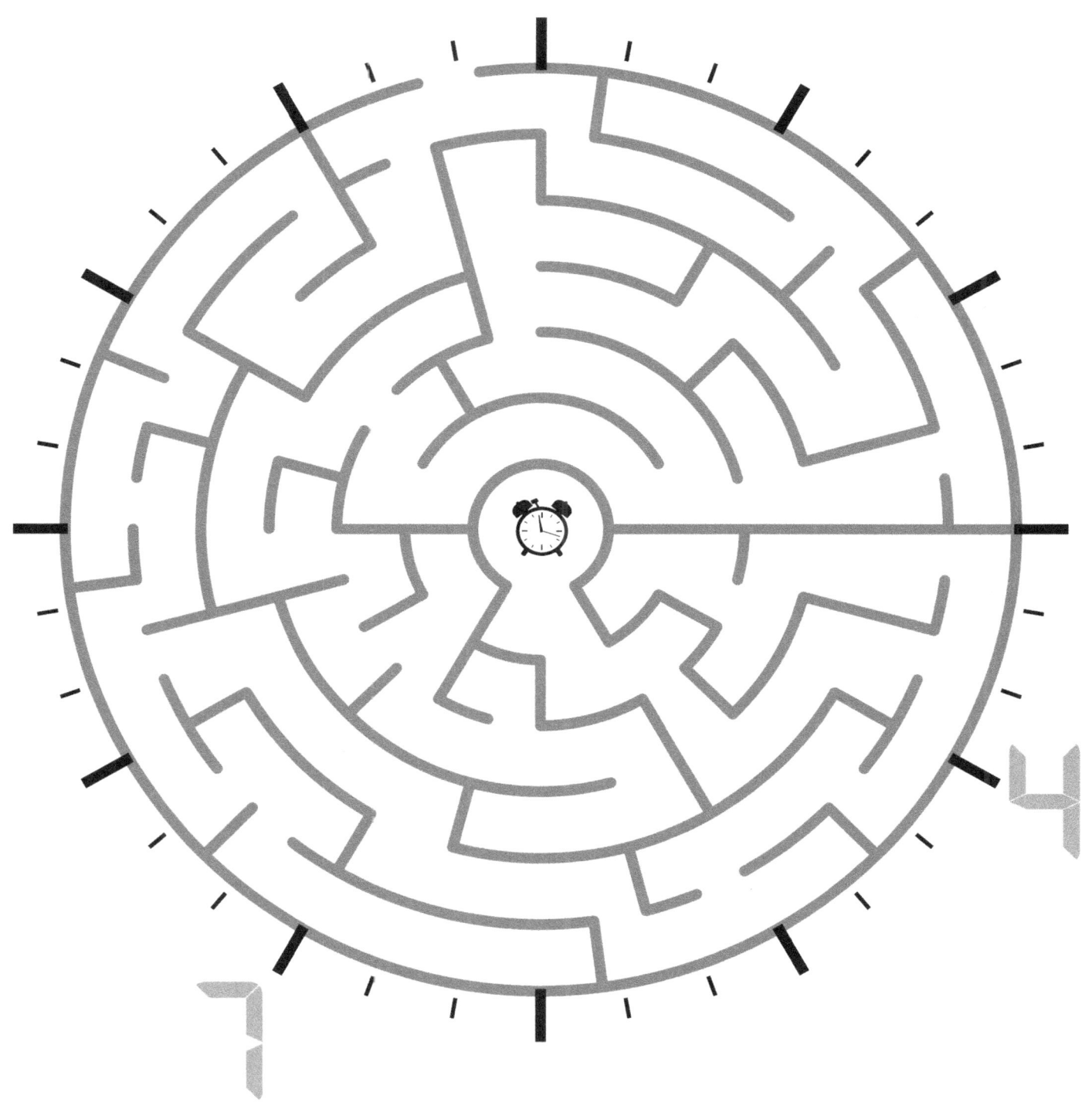

What Time is the Entrance:_____

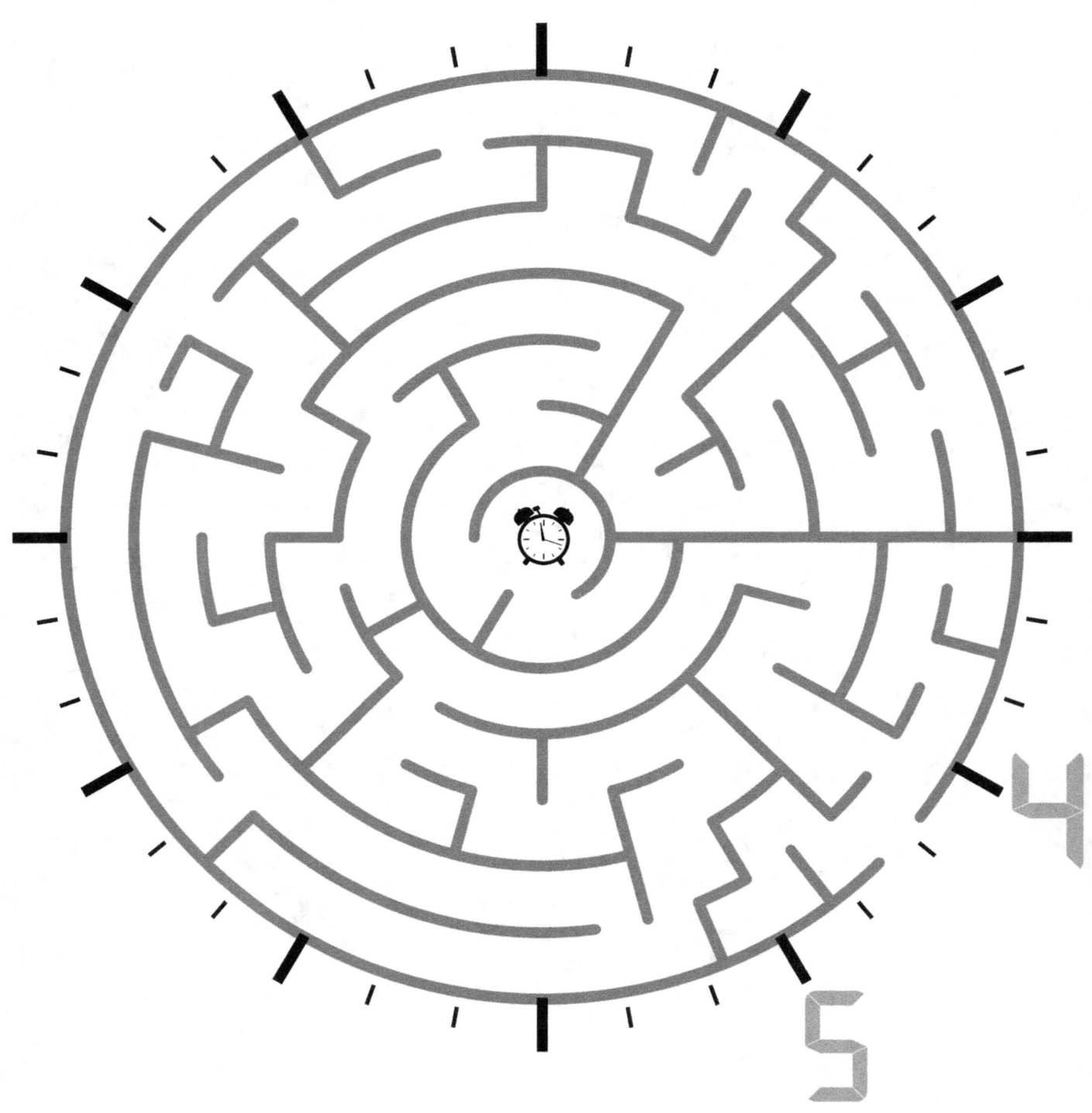

What Time is the Entrance:_____

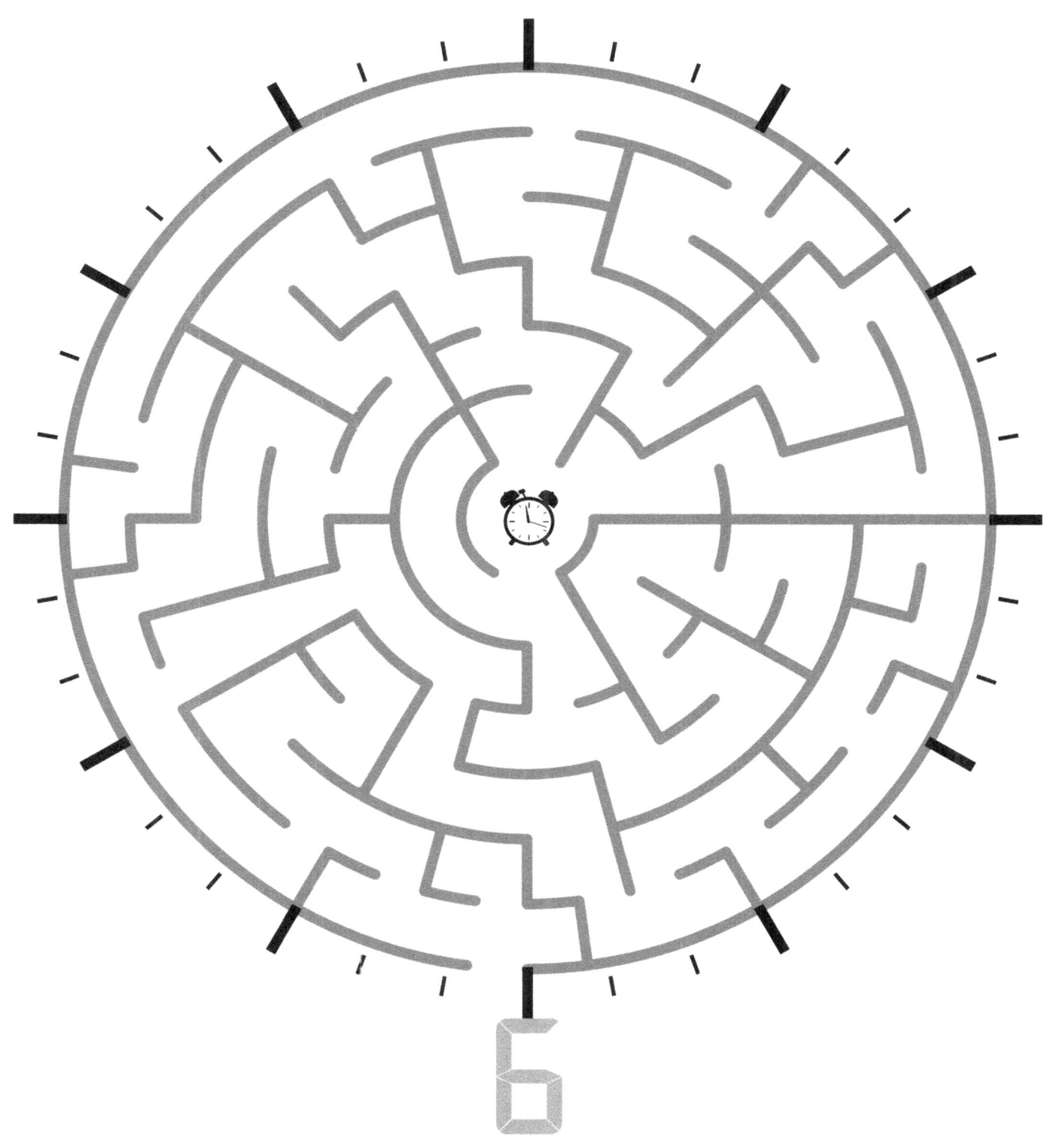

What Time is the Entrance:_____

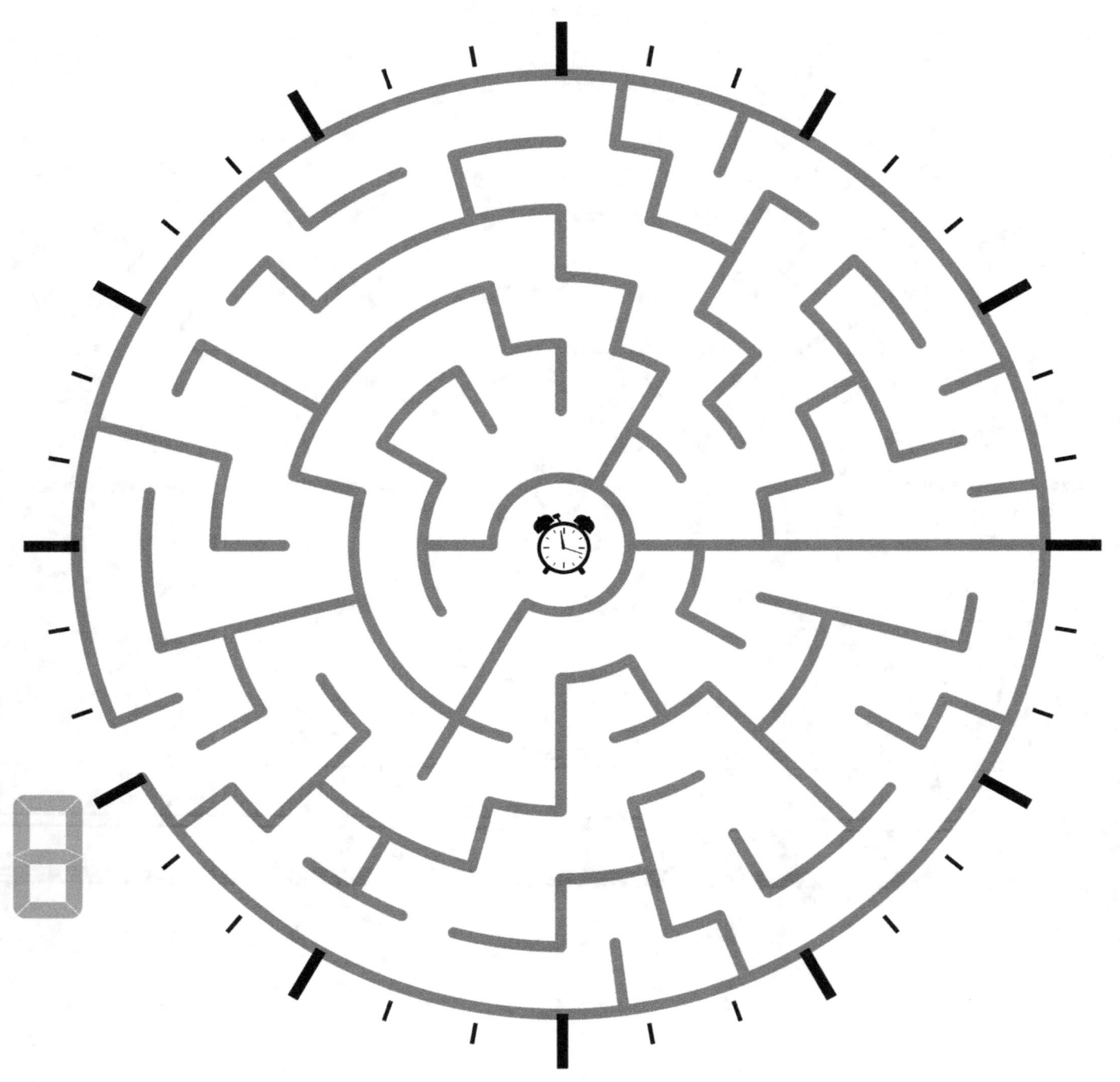

What Time is the Entrance:_____

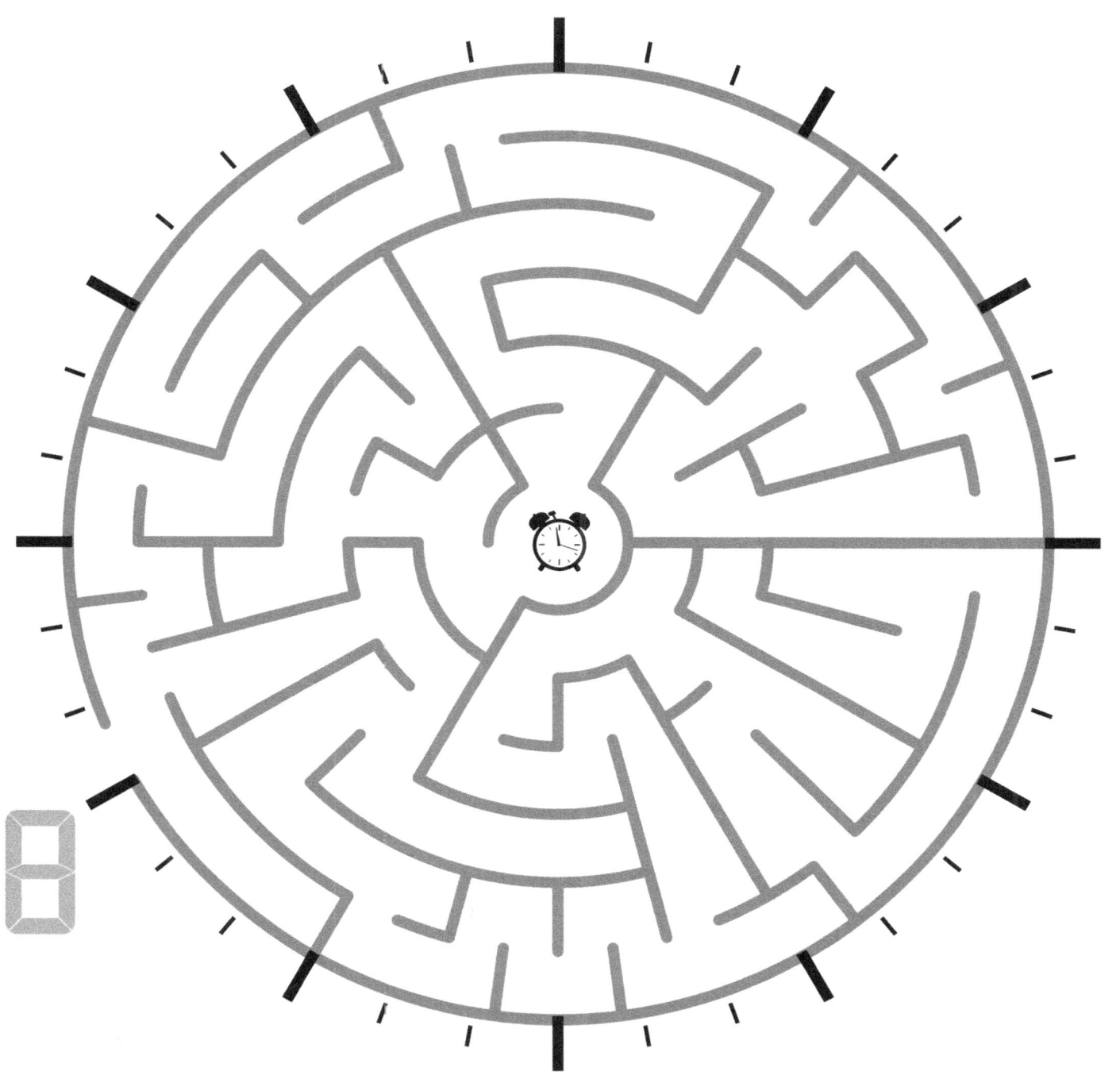

What Time is the Entrance:_____

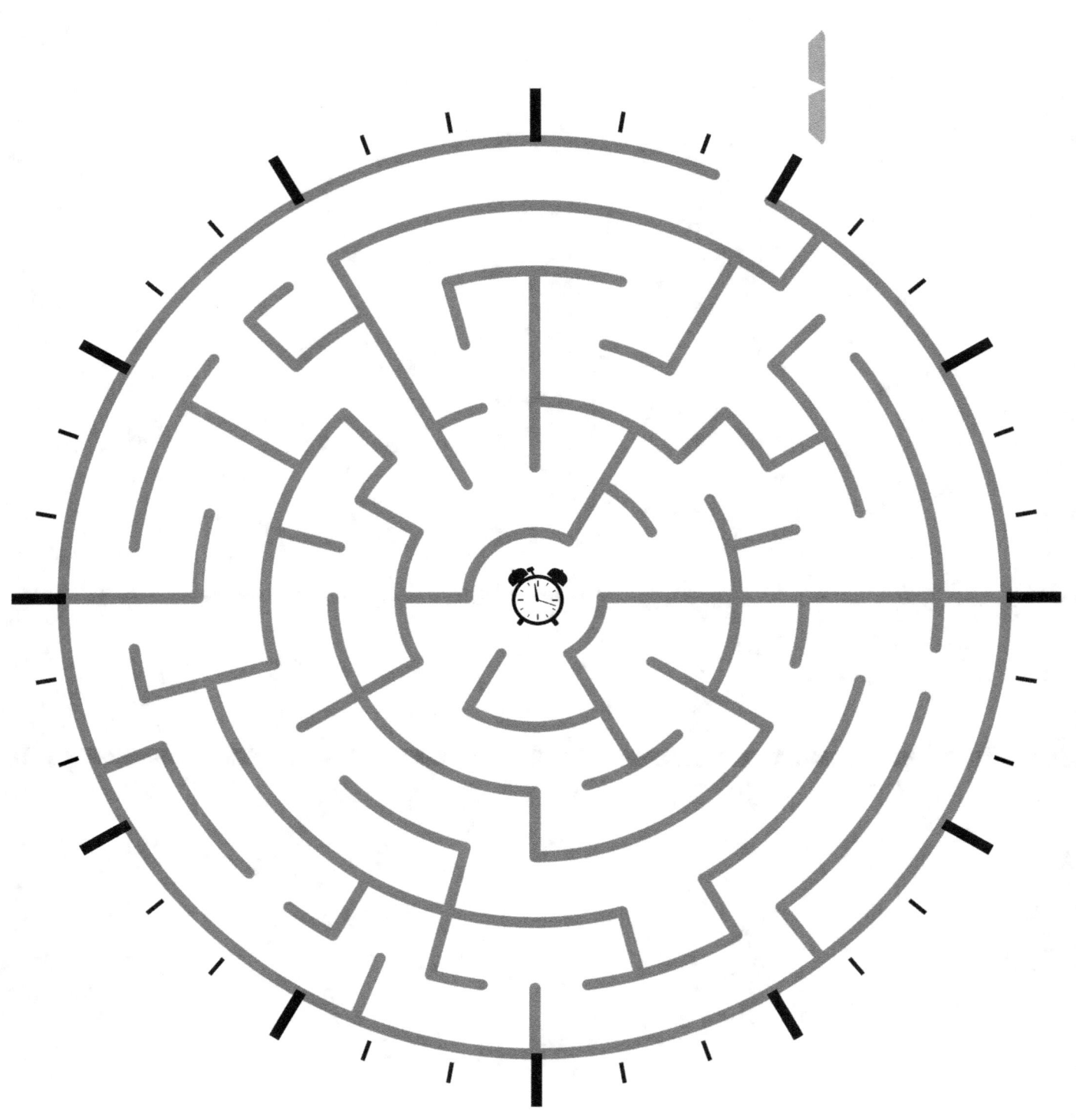

What Time is the Entrance:_____

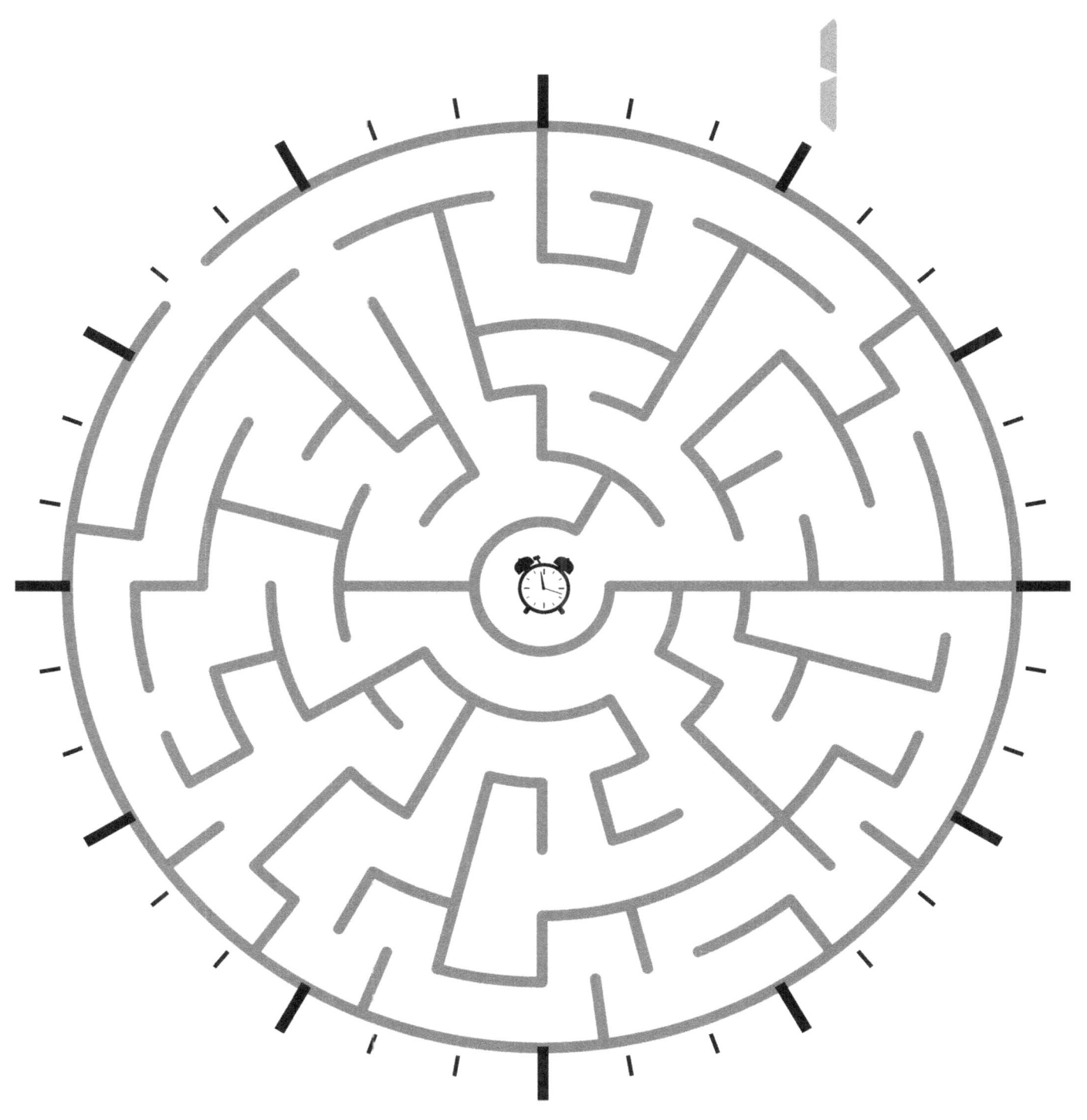

What Time is the Entrance:_____

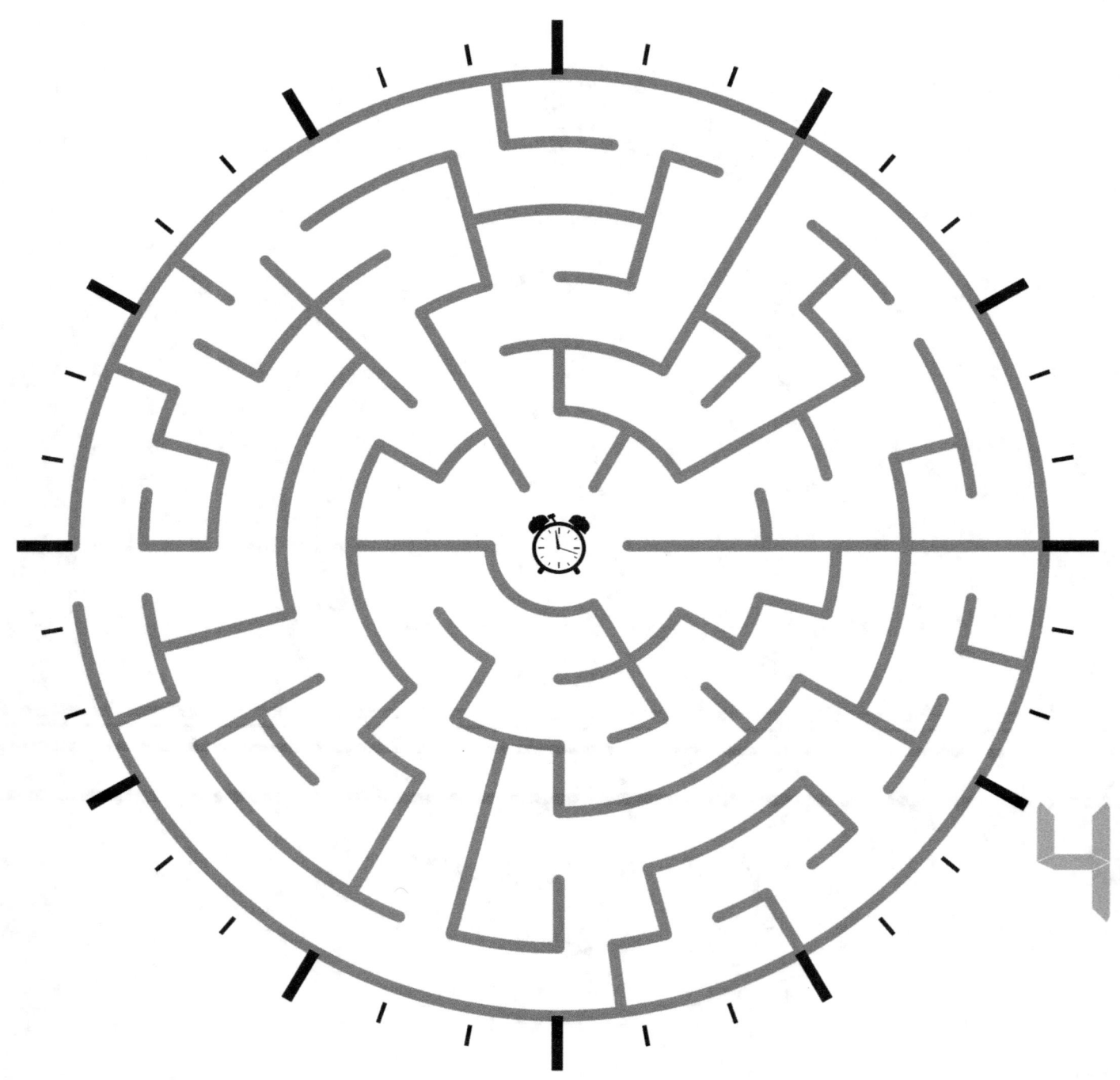

What Time is the Entrance:_____

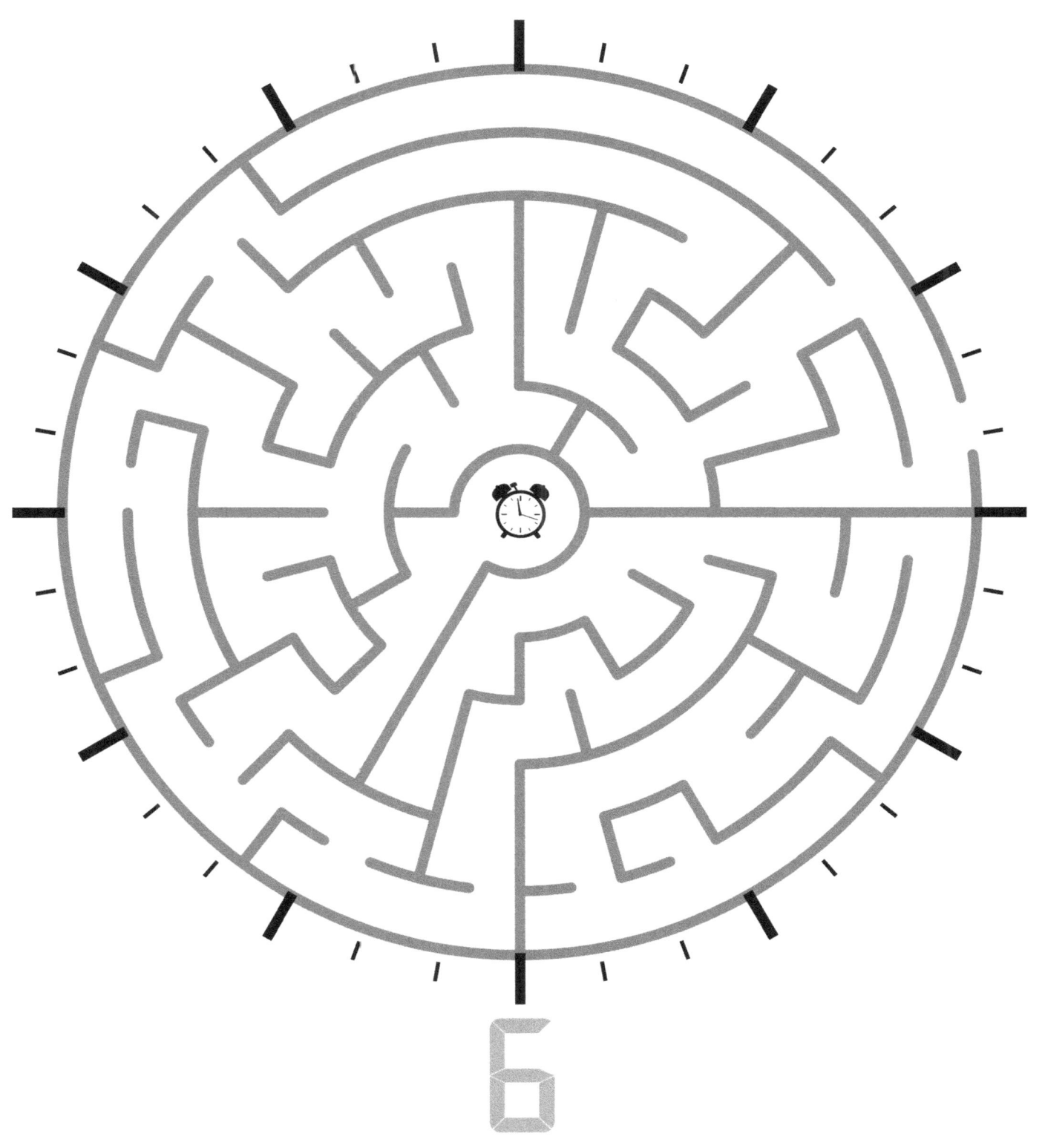

What Time is the Entrance:_____

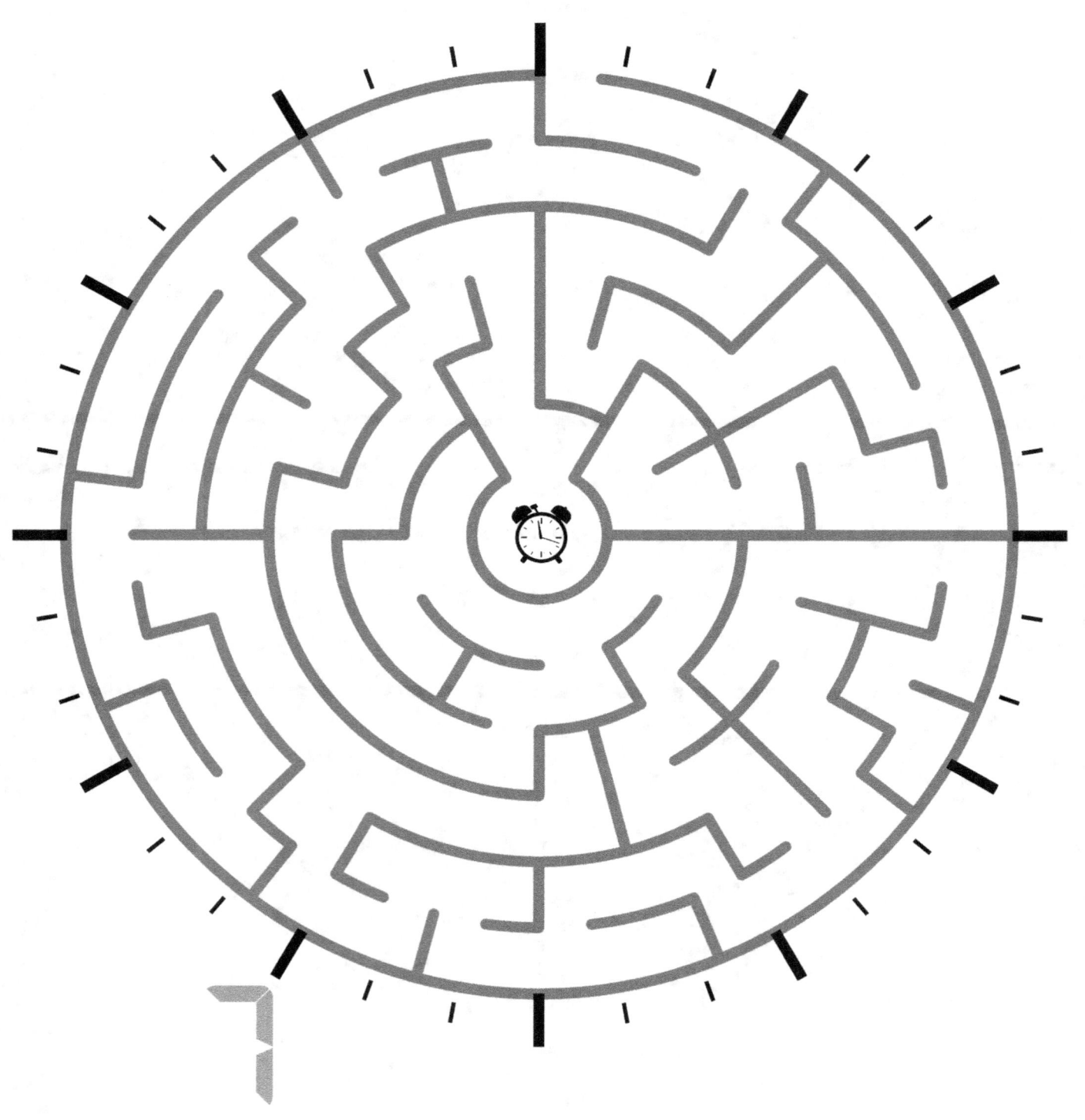

What Time is the Entrance:_____

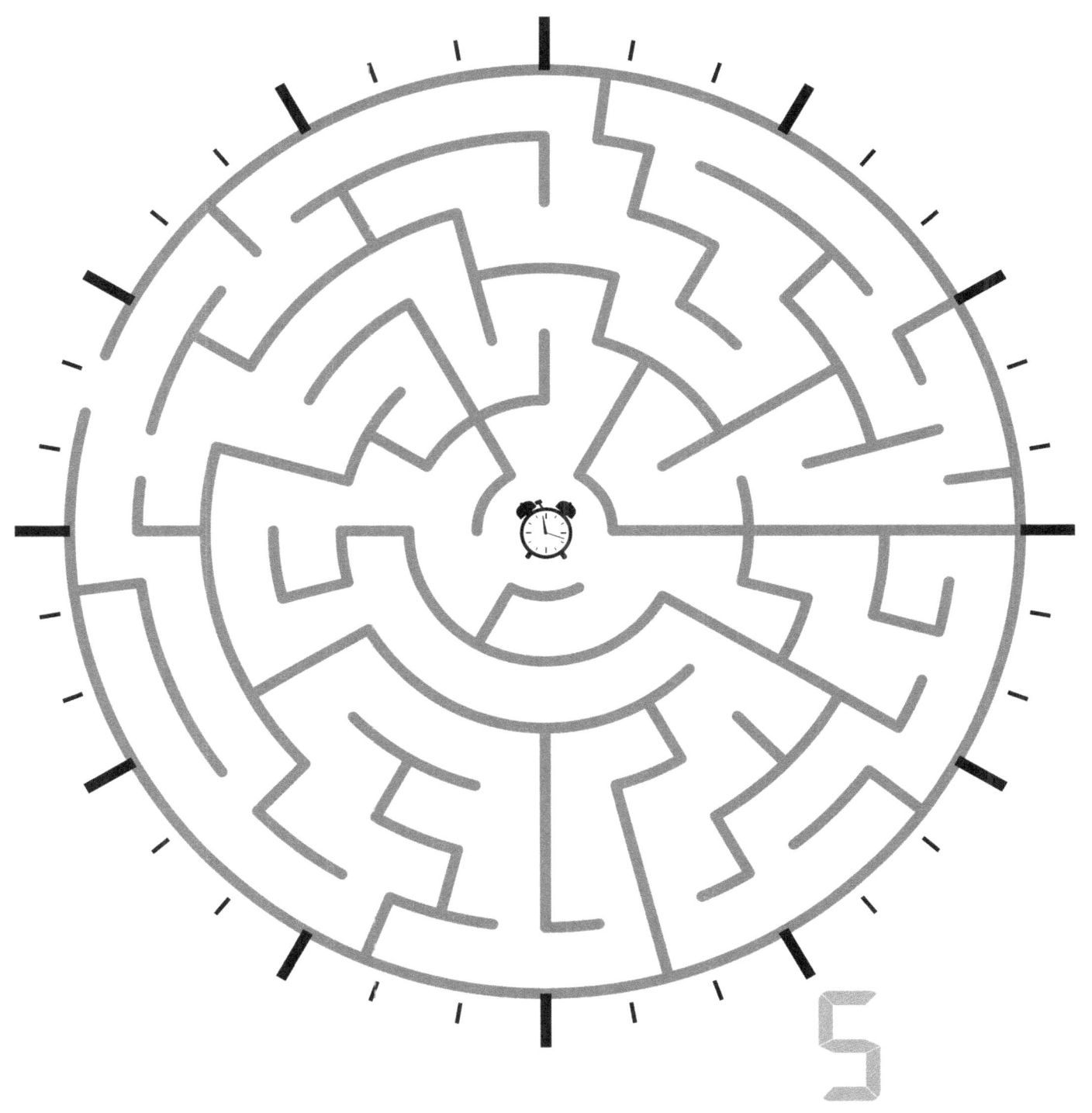

What Time is the Entrance:_____

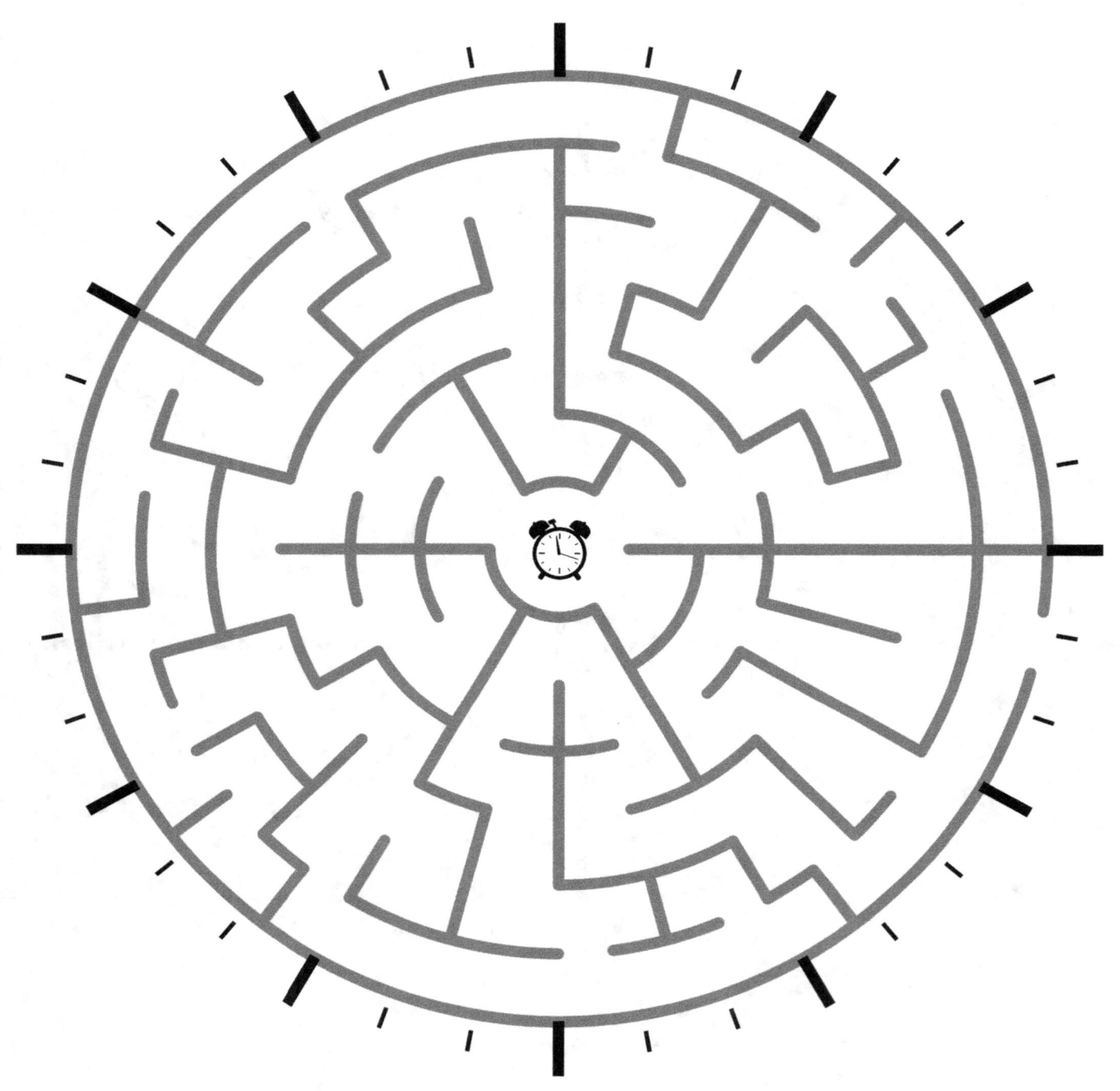

What Time is the Entrance:_____

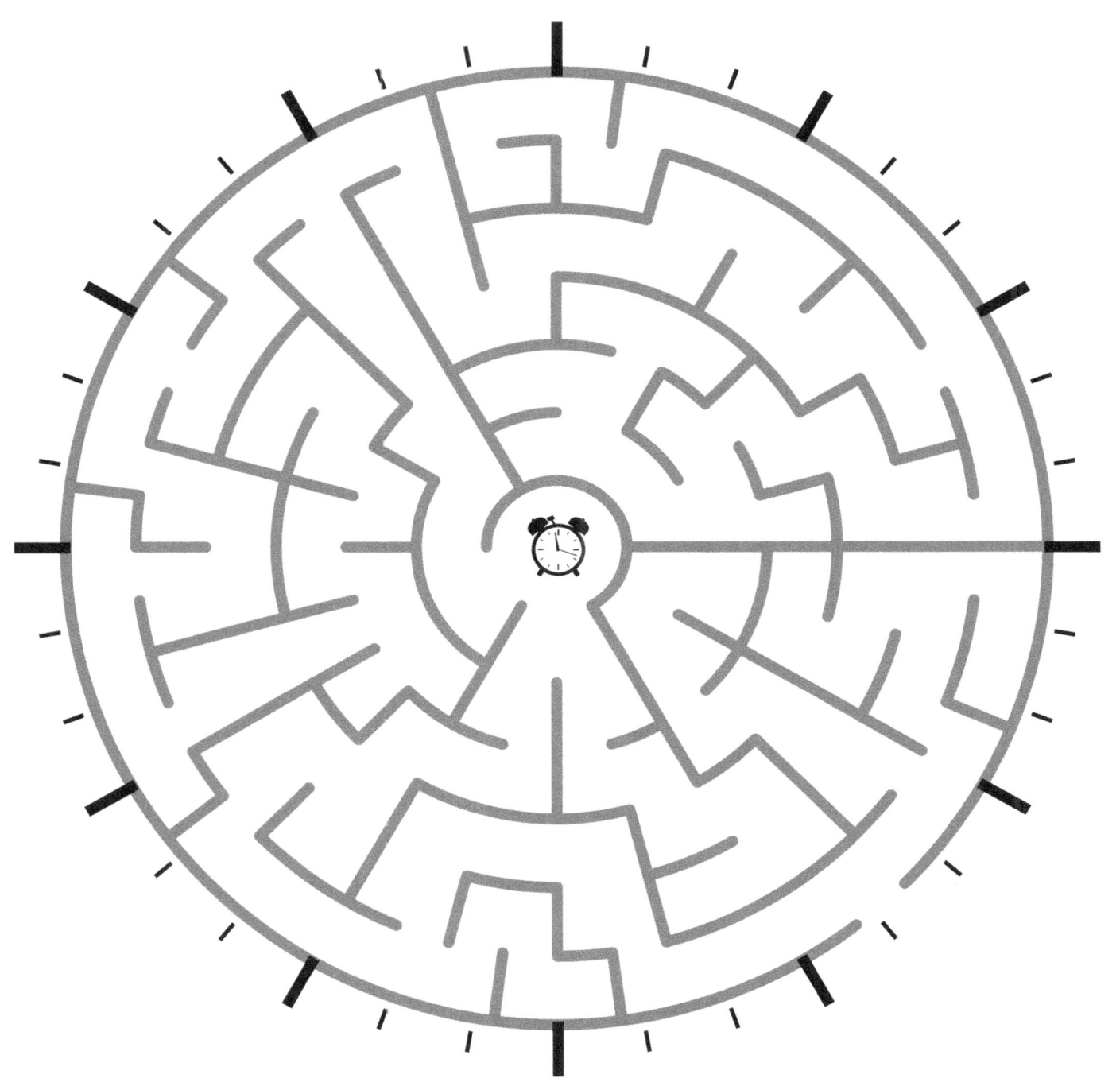

What Time is the Entrance:_____

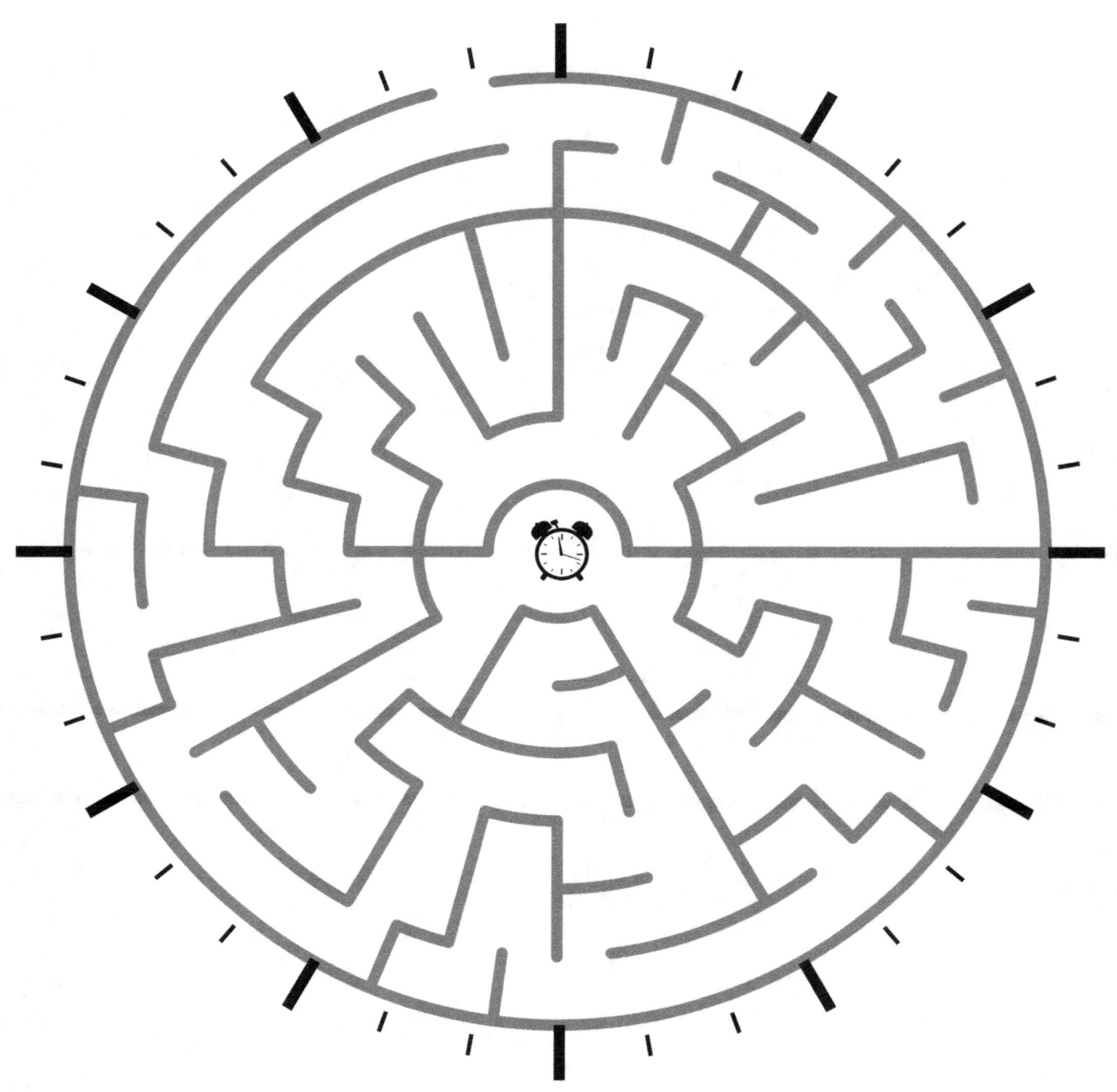

What Time is the Entrance:_____

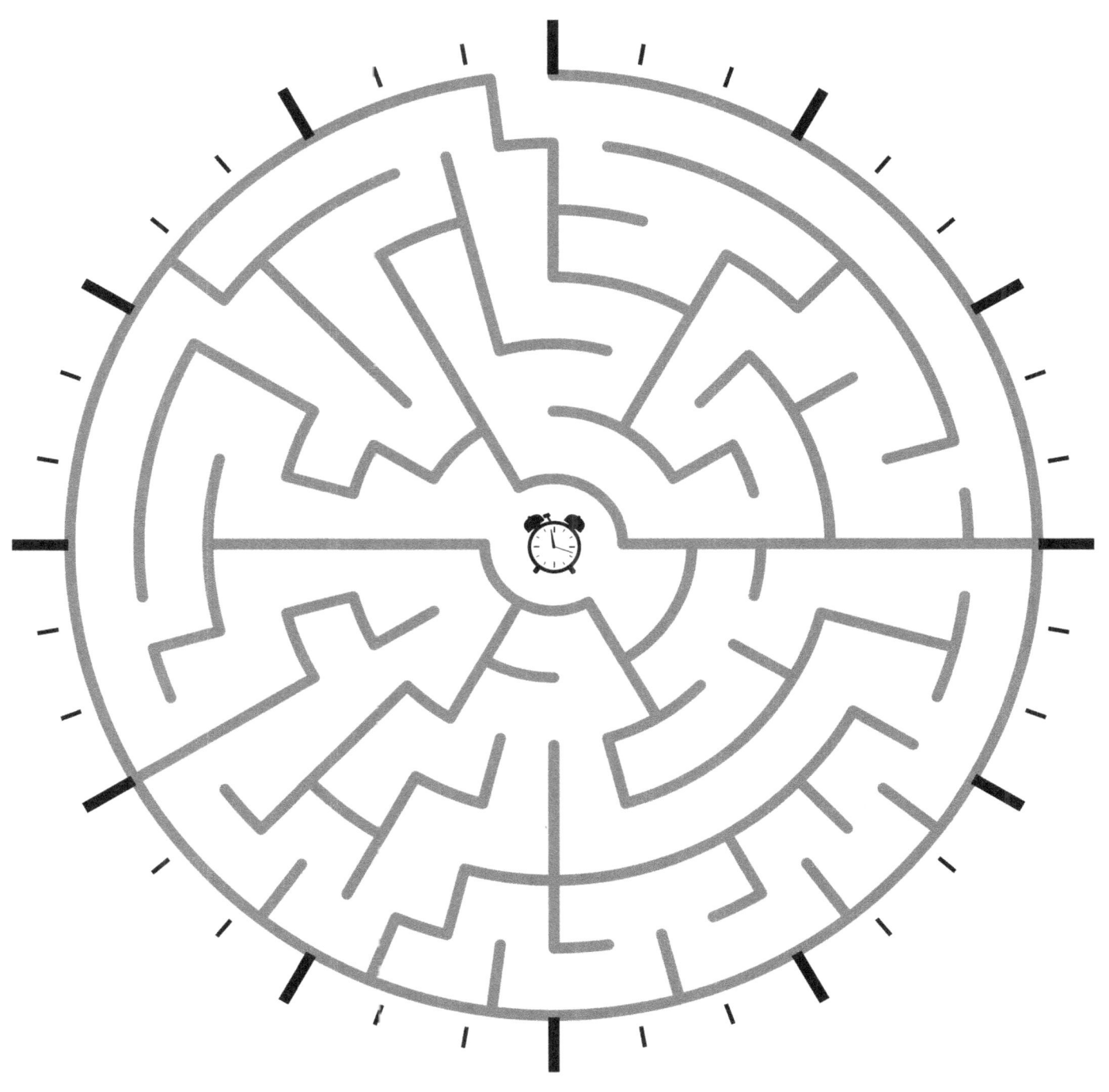

What Time is the Entrance:_____

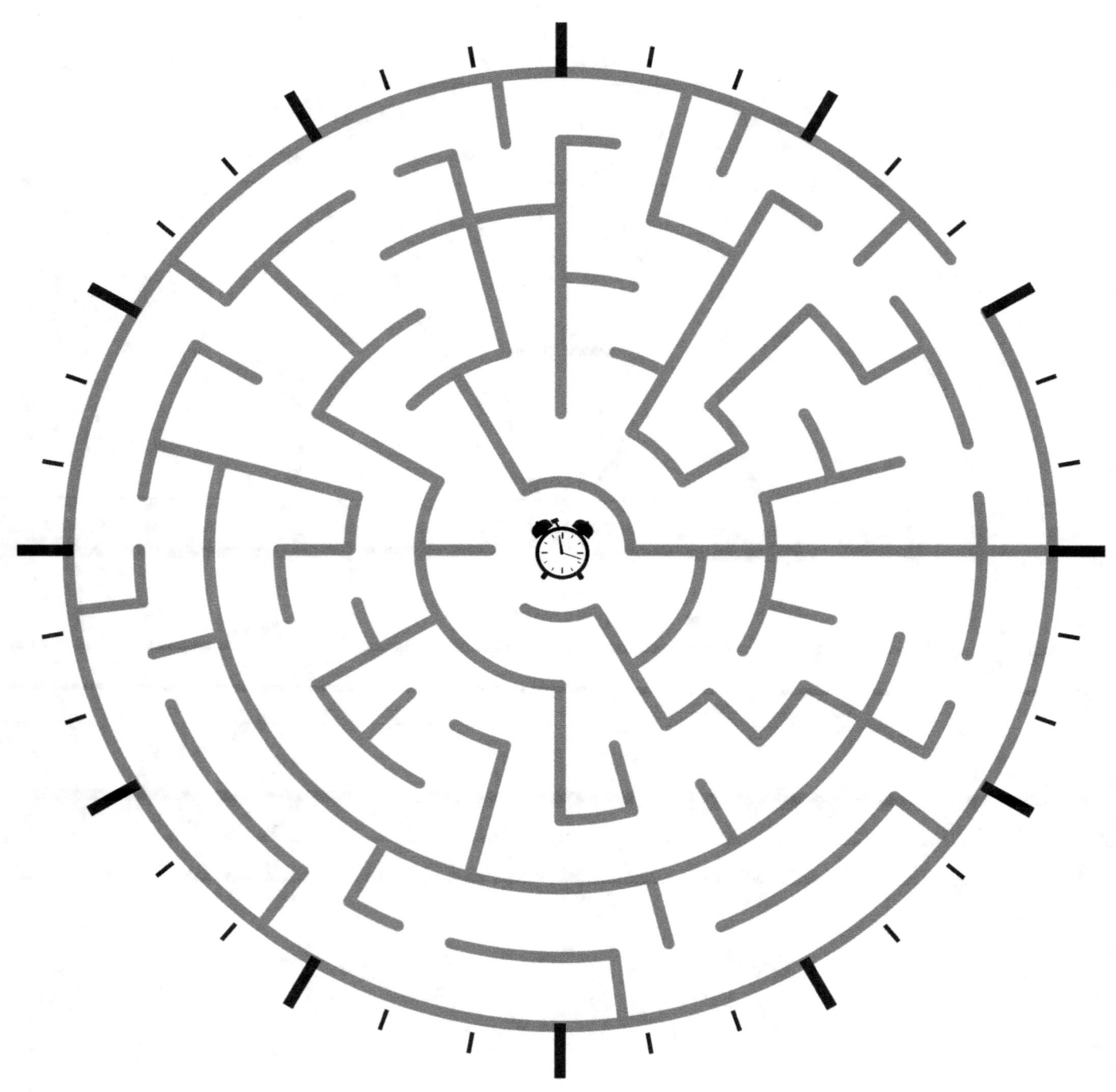

What Time is the Entrance:_____

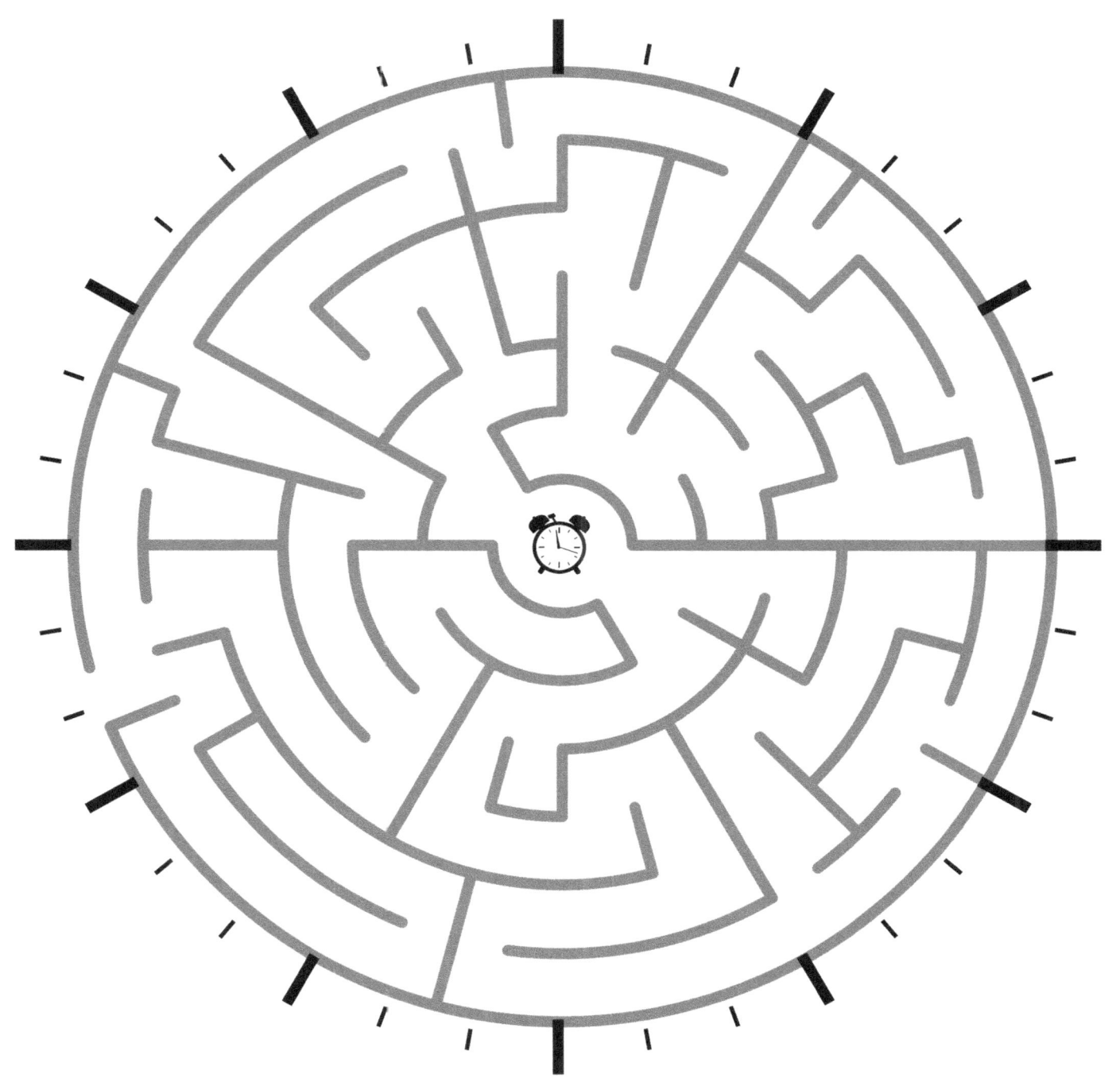

What Time is the Entrance:_____

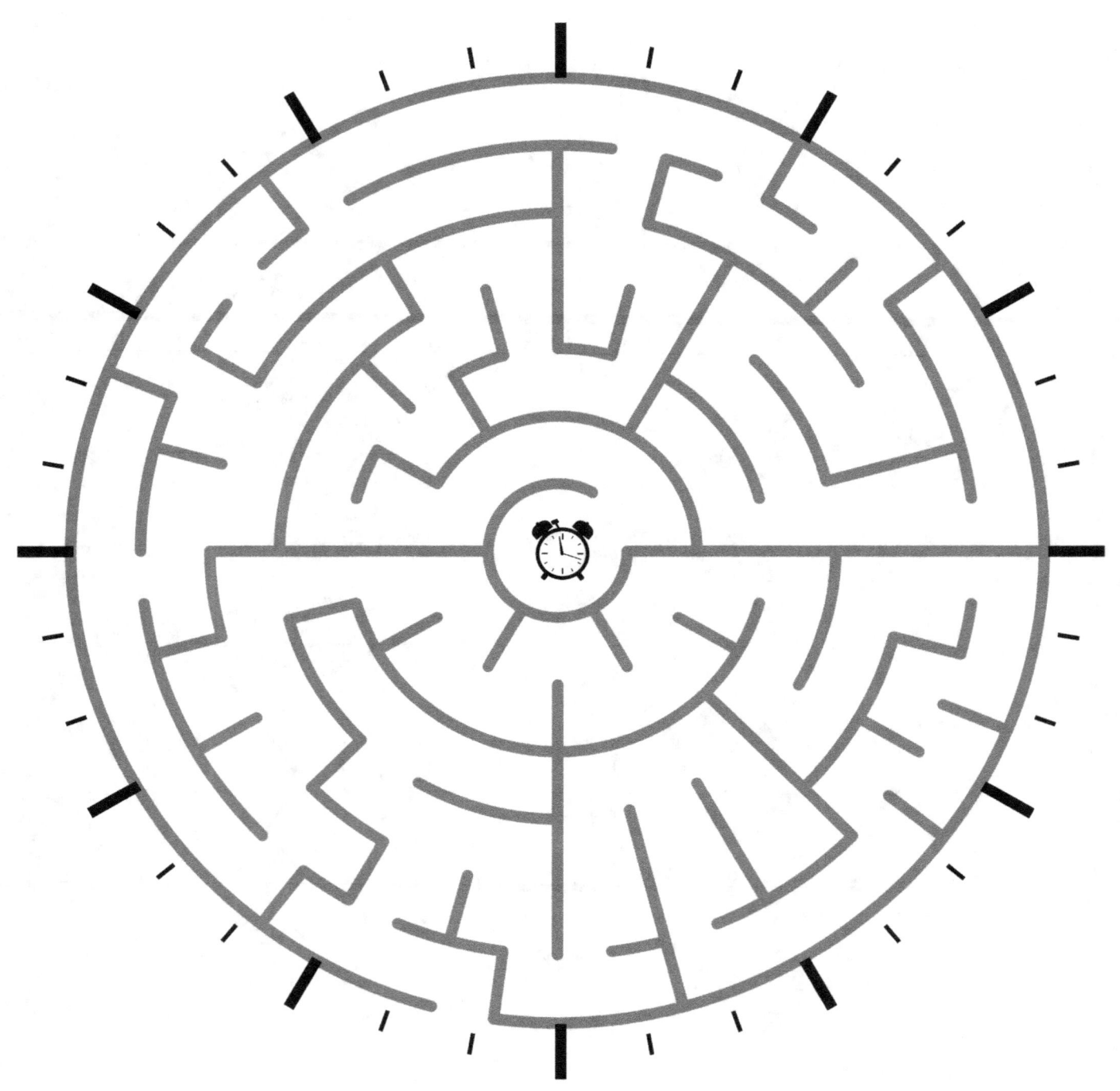

What Time is the Entrance:_____

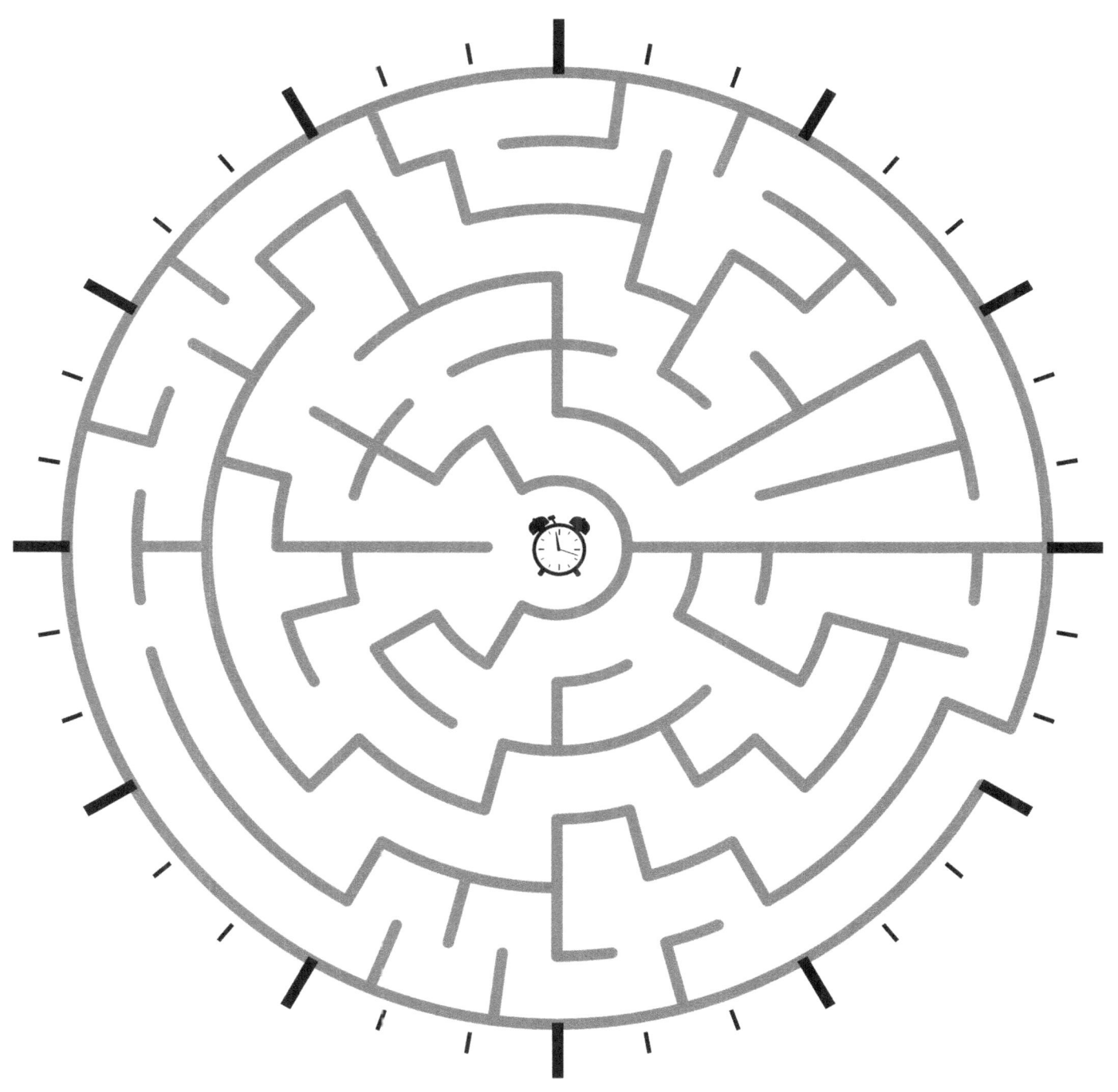

What Time is the Entrance:_____

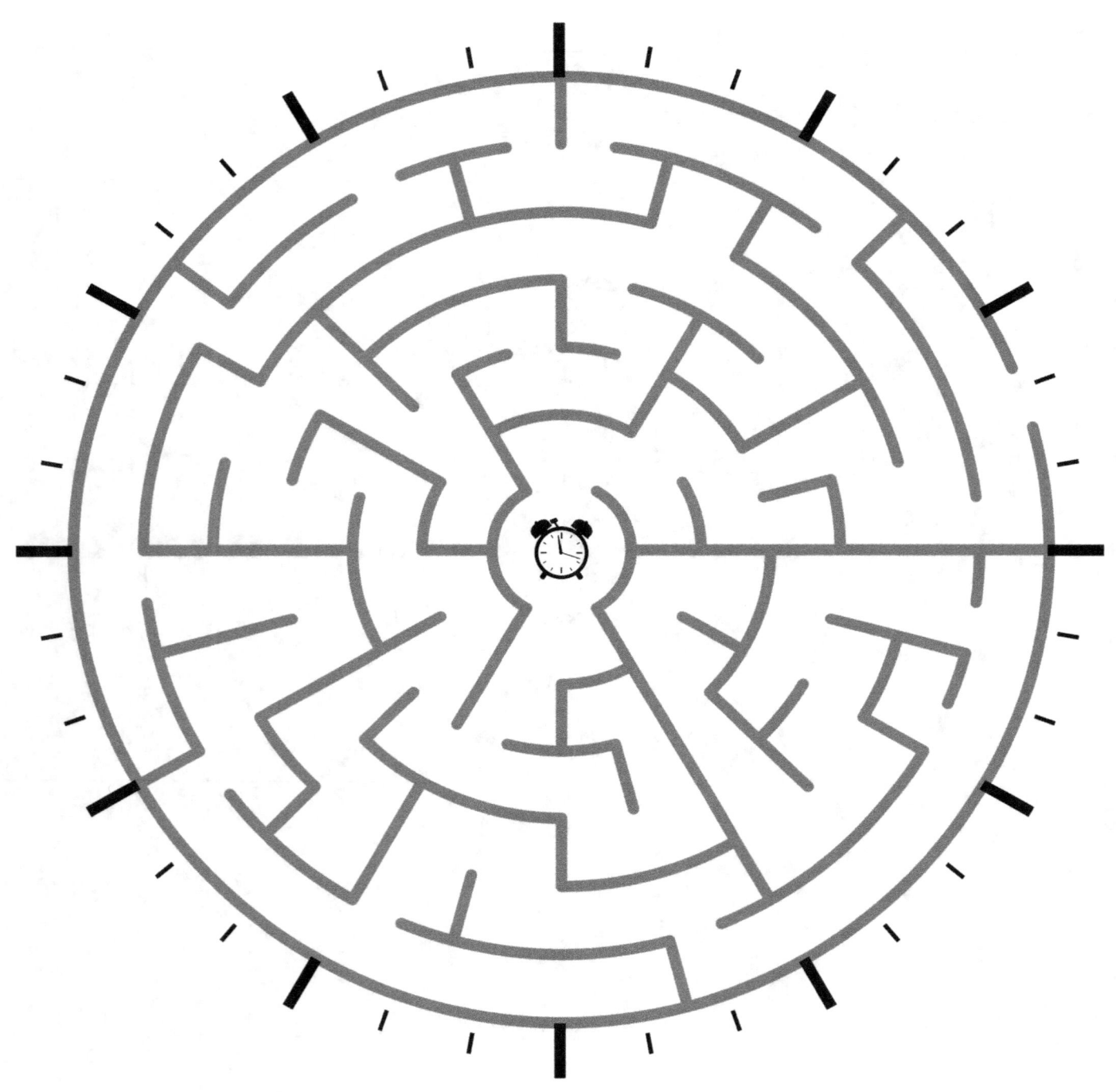

What Time is the Entrance:_____

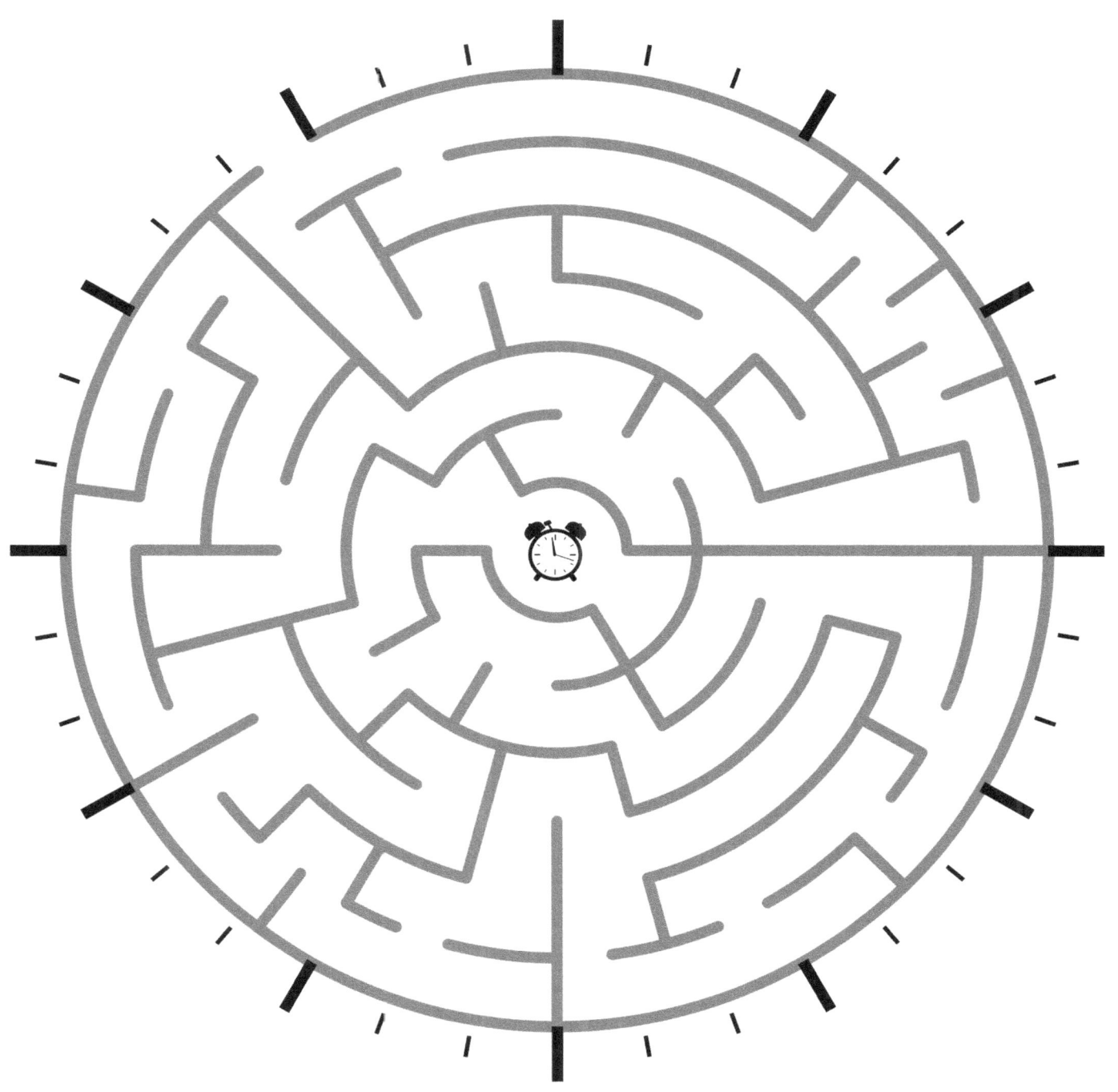

Page 46

What Time is the Entrance:_____

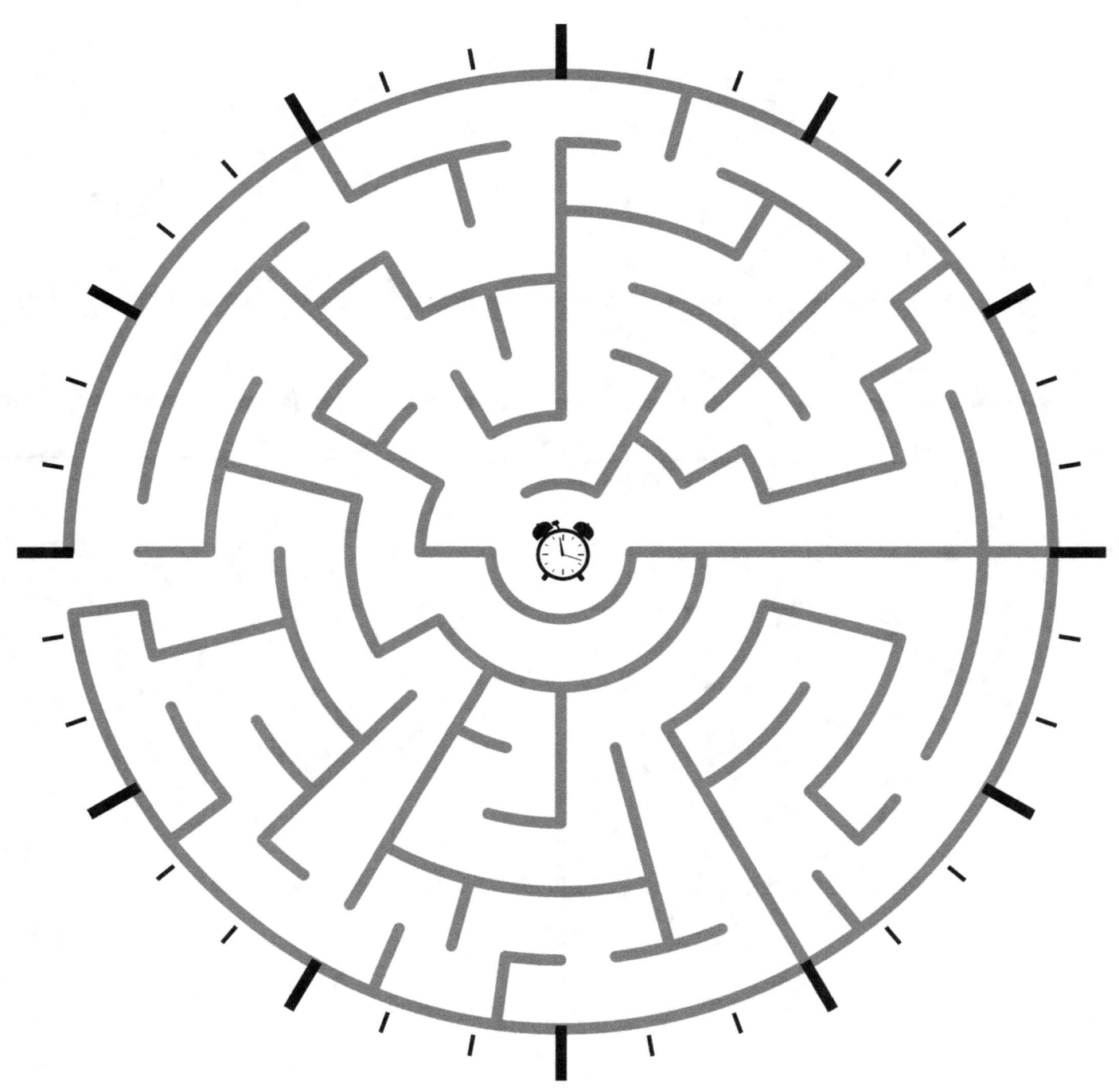

What Time is the Entrance:_____

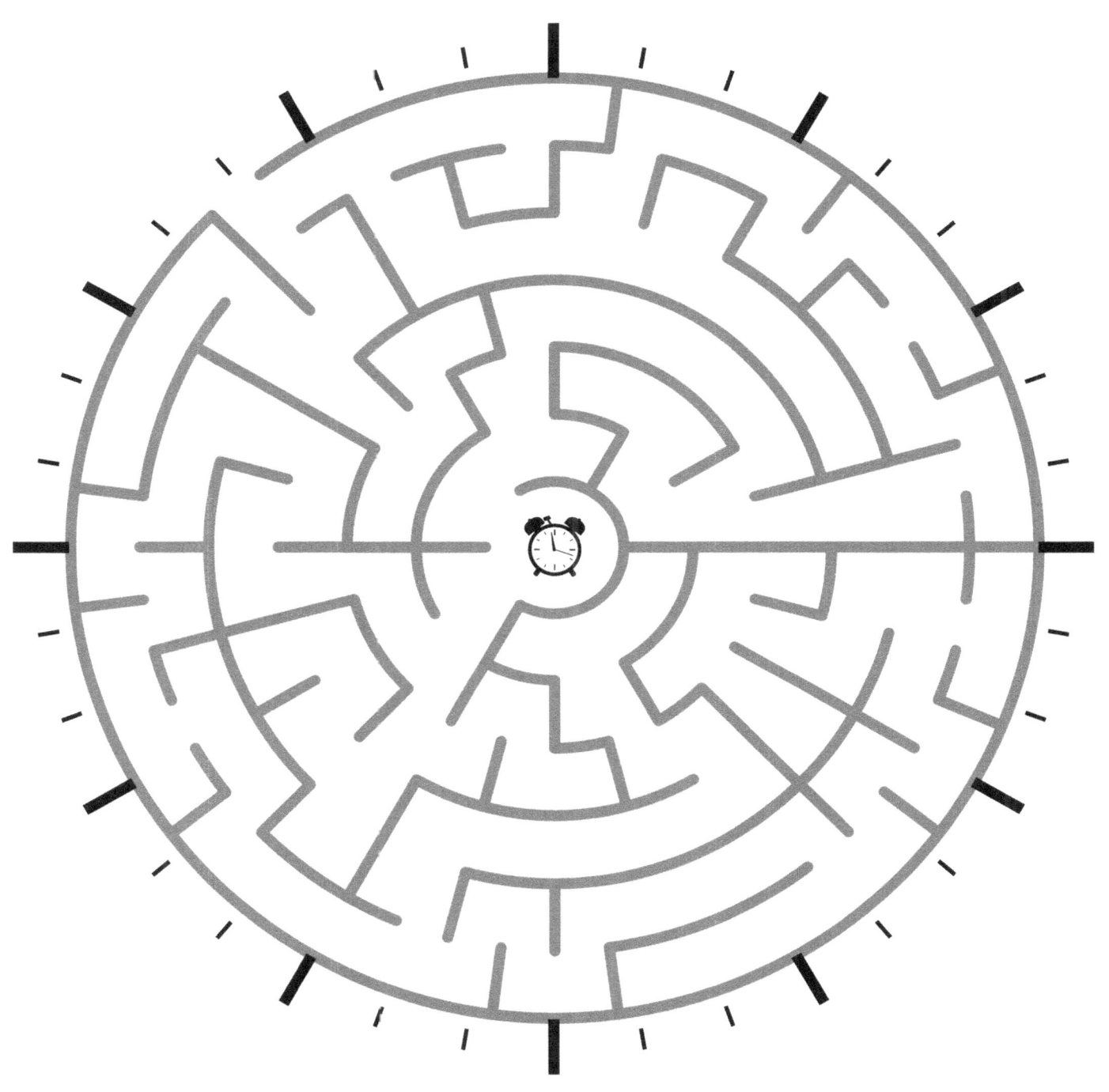

What Time is the Entrance:_____

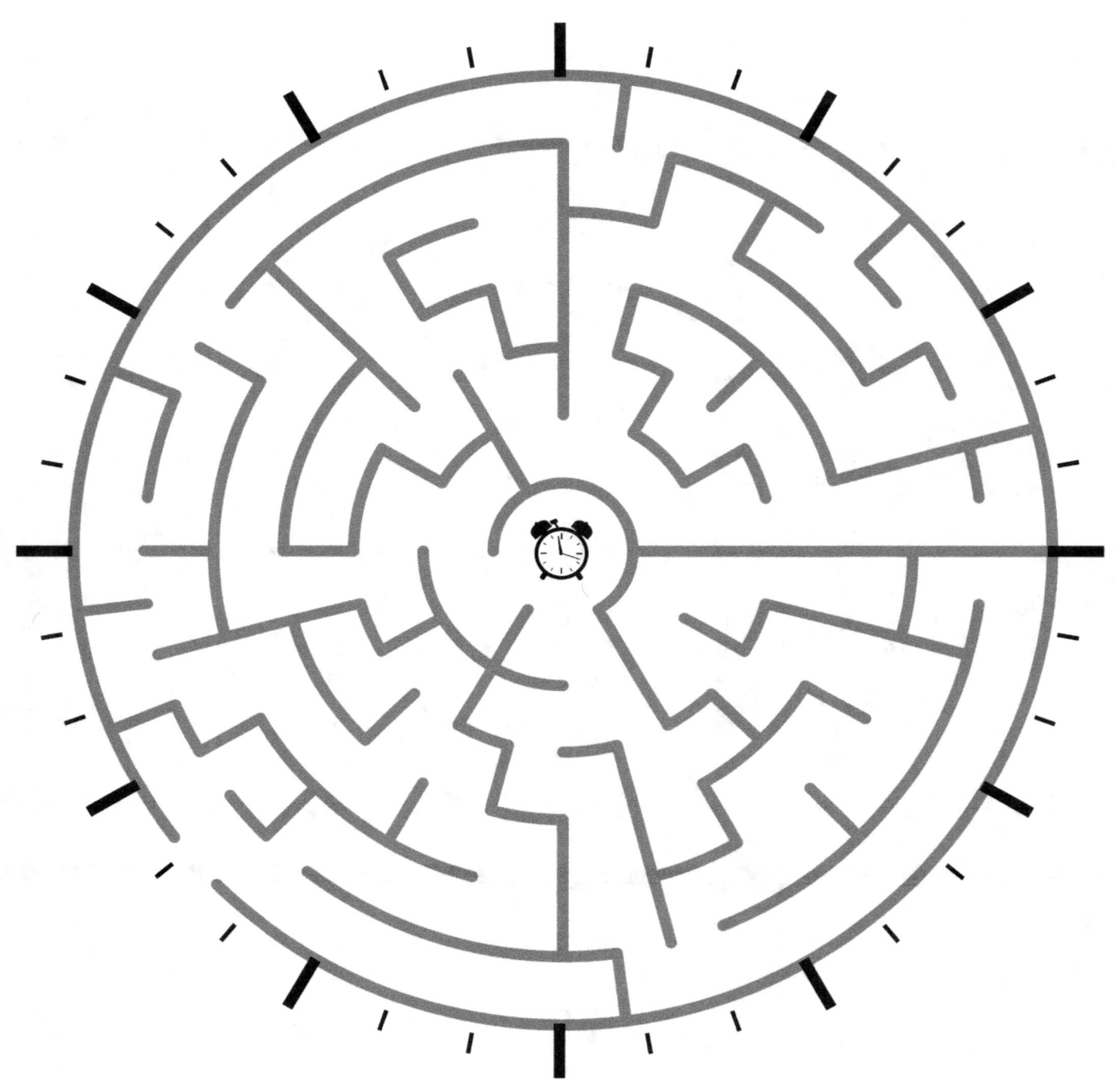

What Time is the Entrance:_____

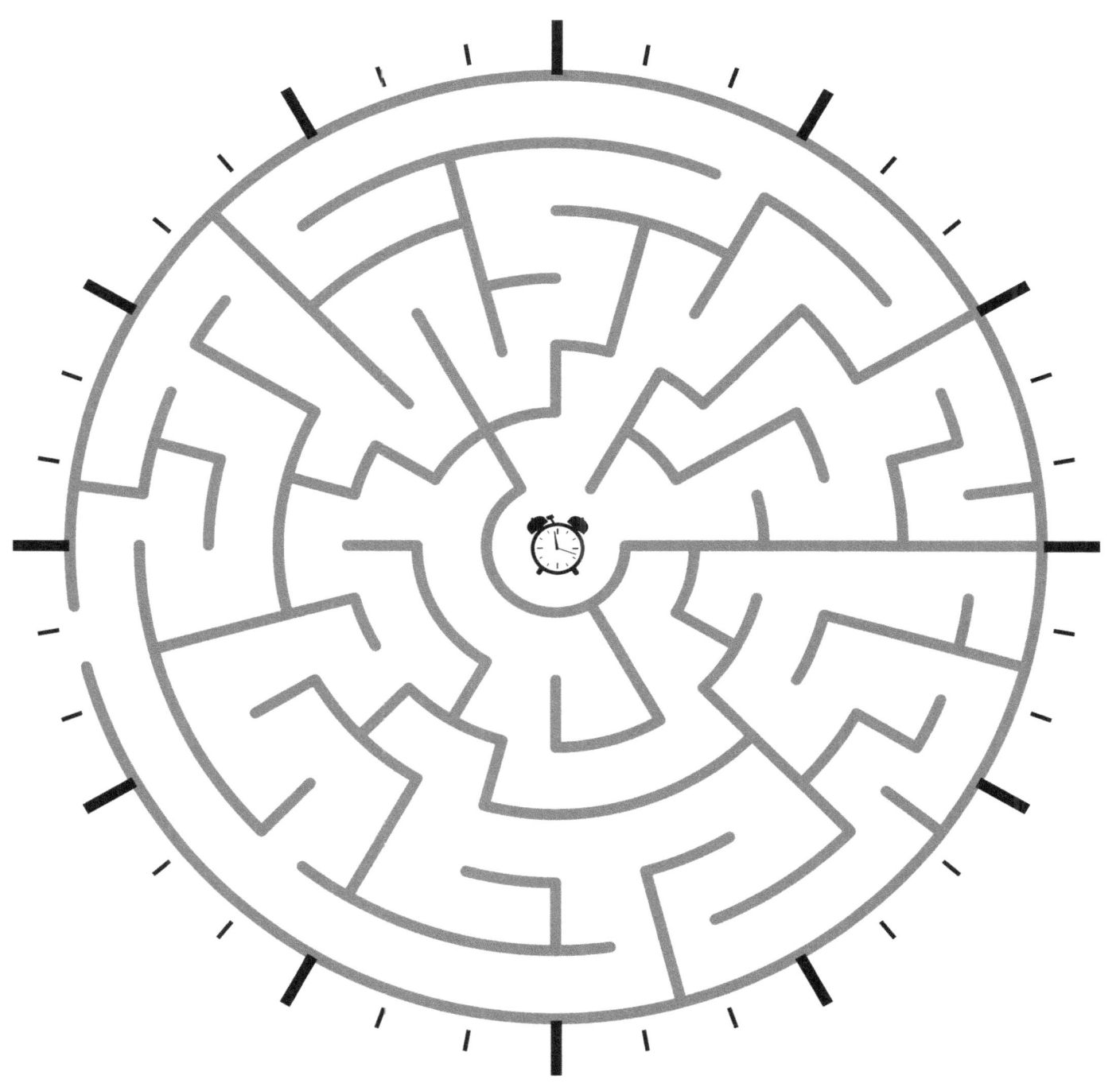

What Time is the Entrance:_____

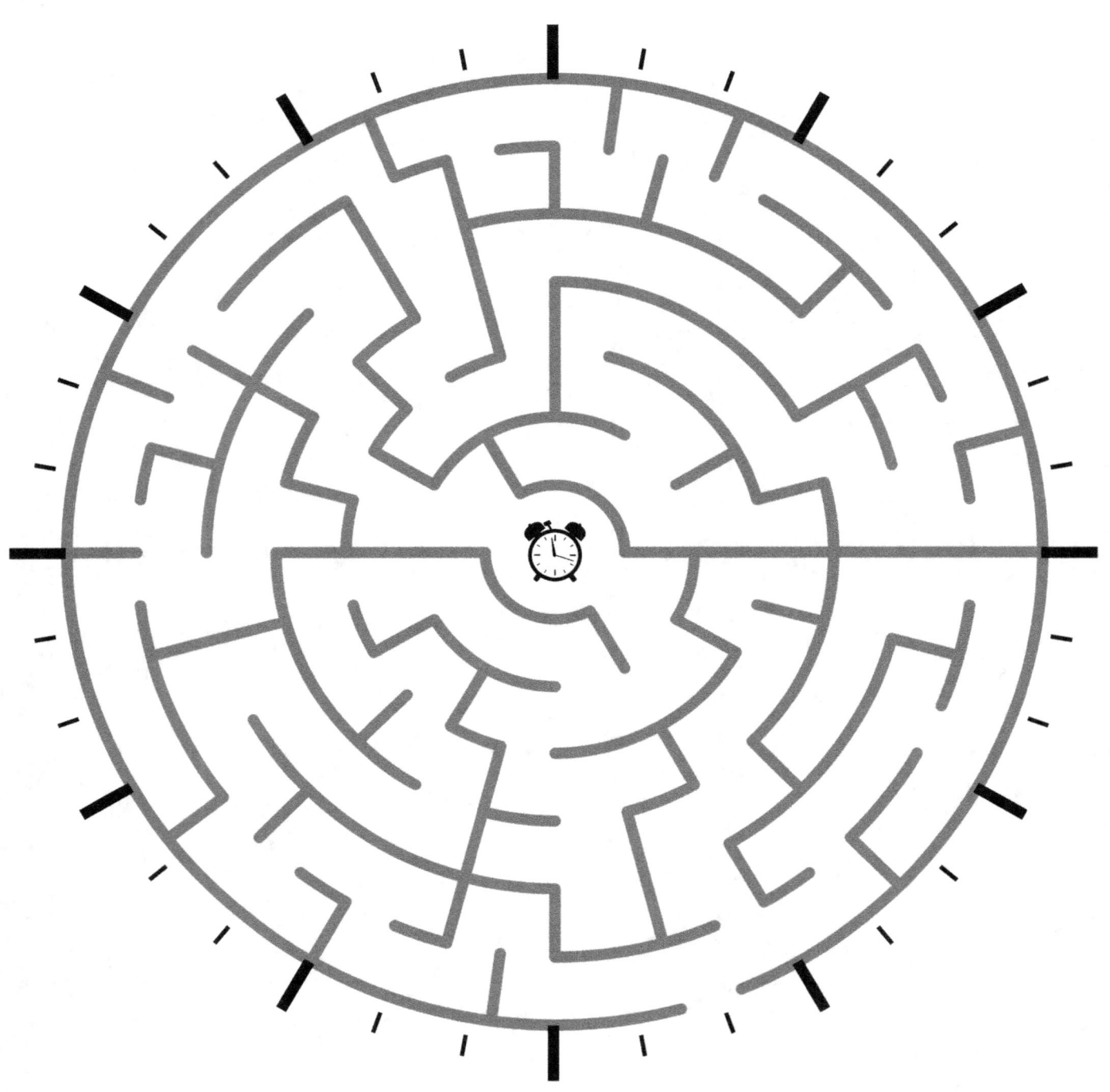

What Time is the Entrance:_____

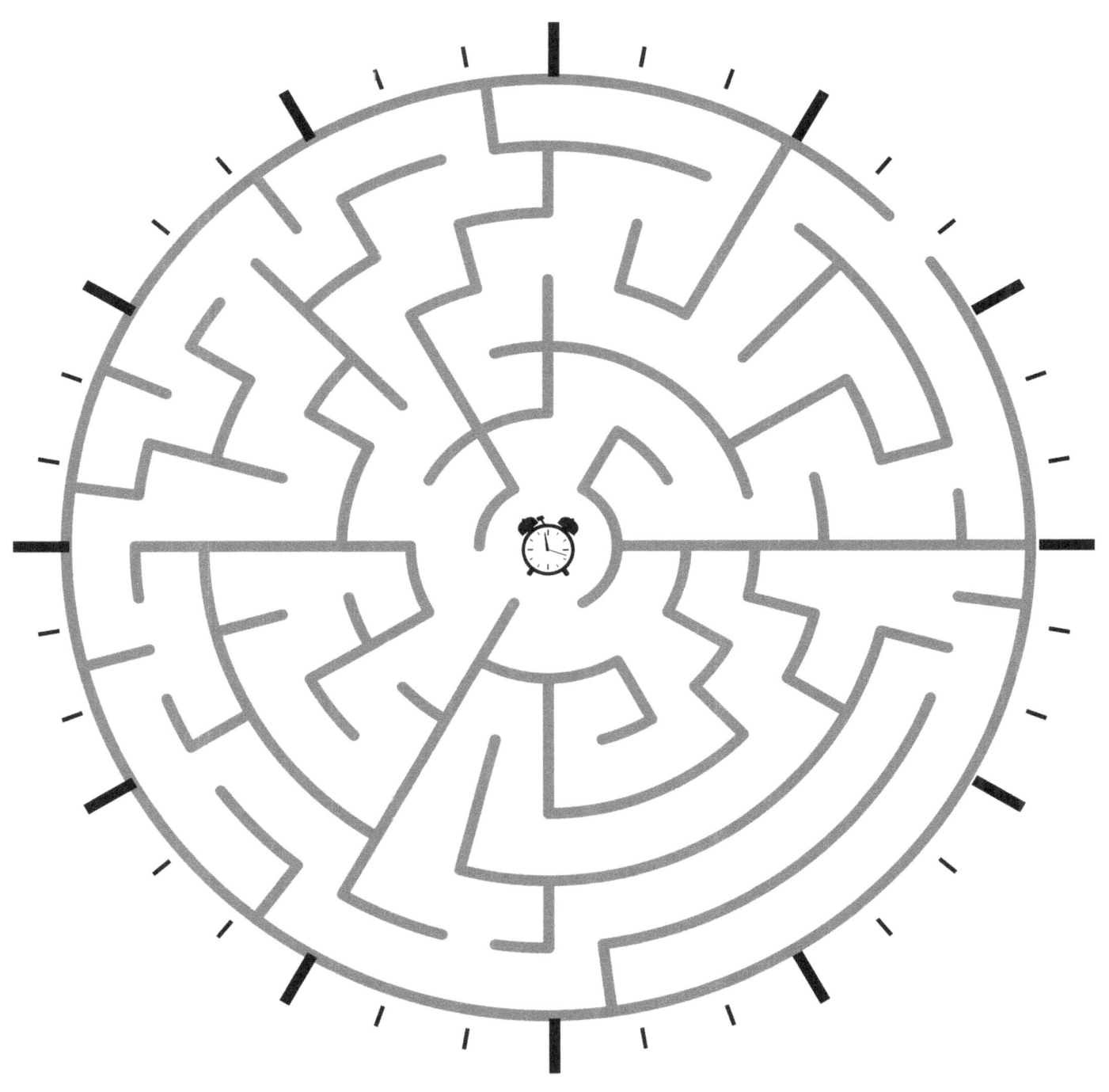

What Time is the Entrance:_____

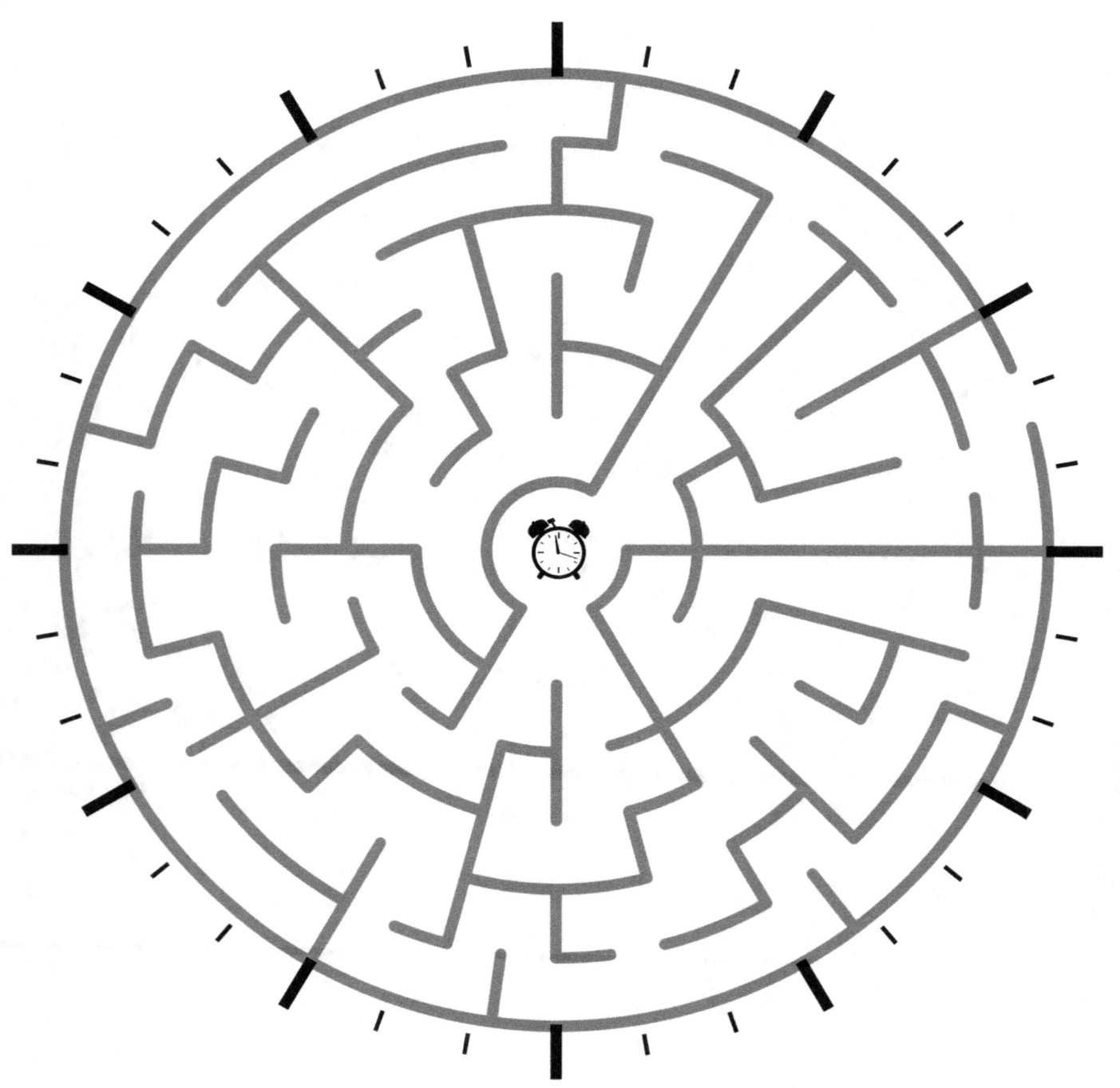

What Time is the Entrance:_____

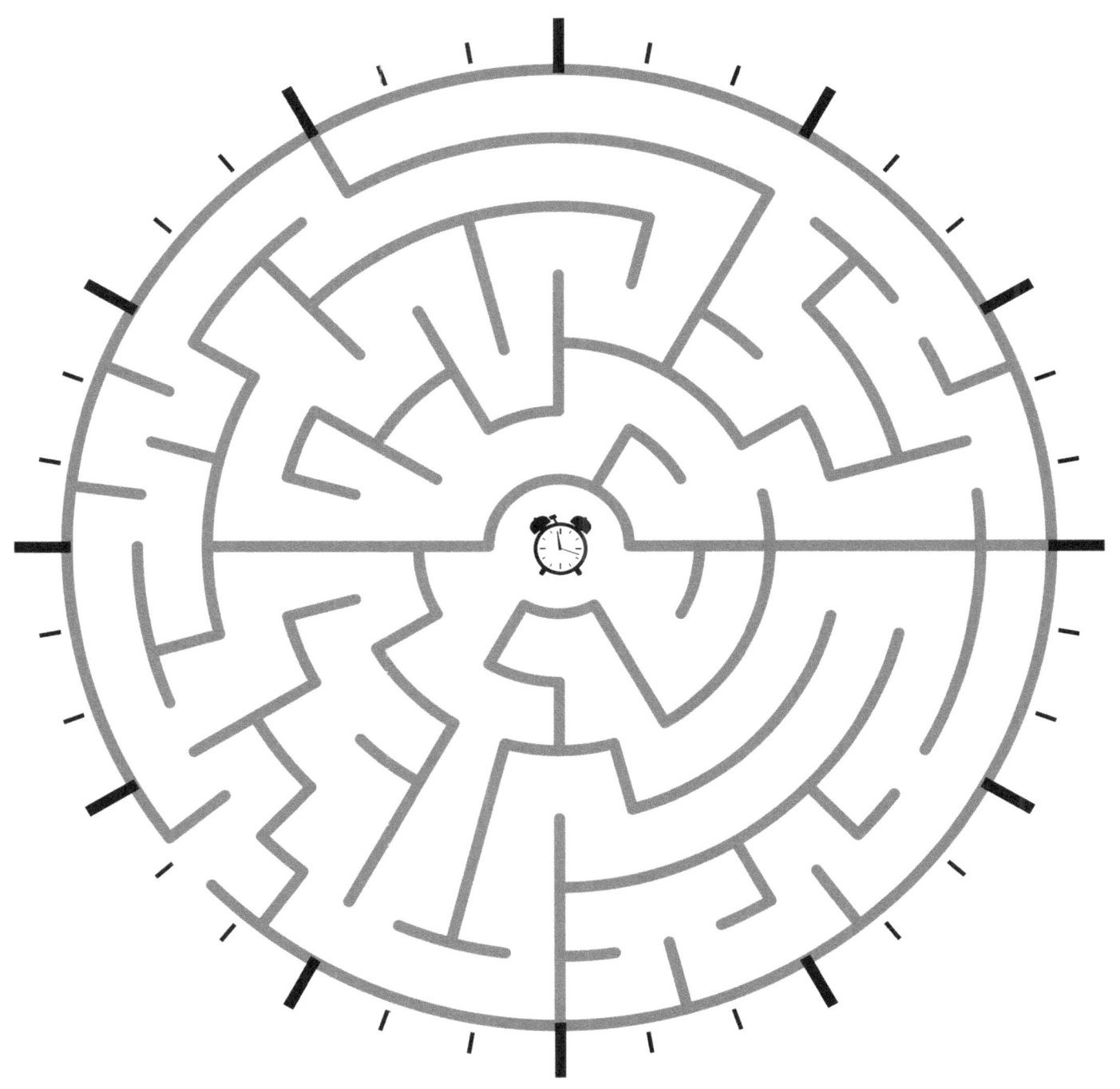

What Time is the Entrance:_____

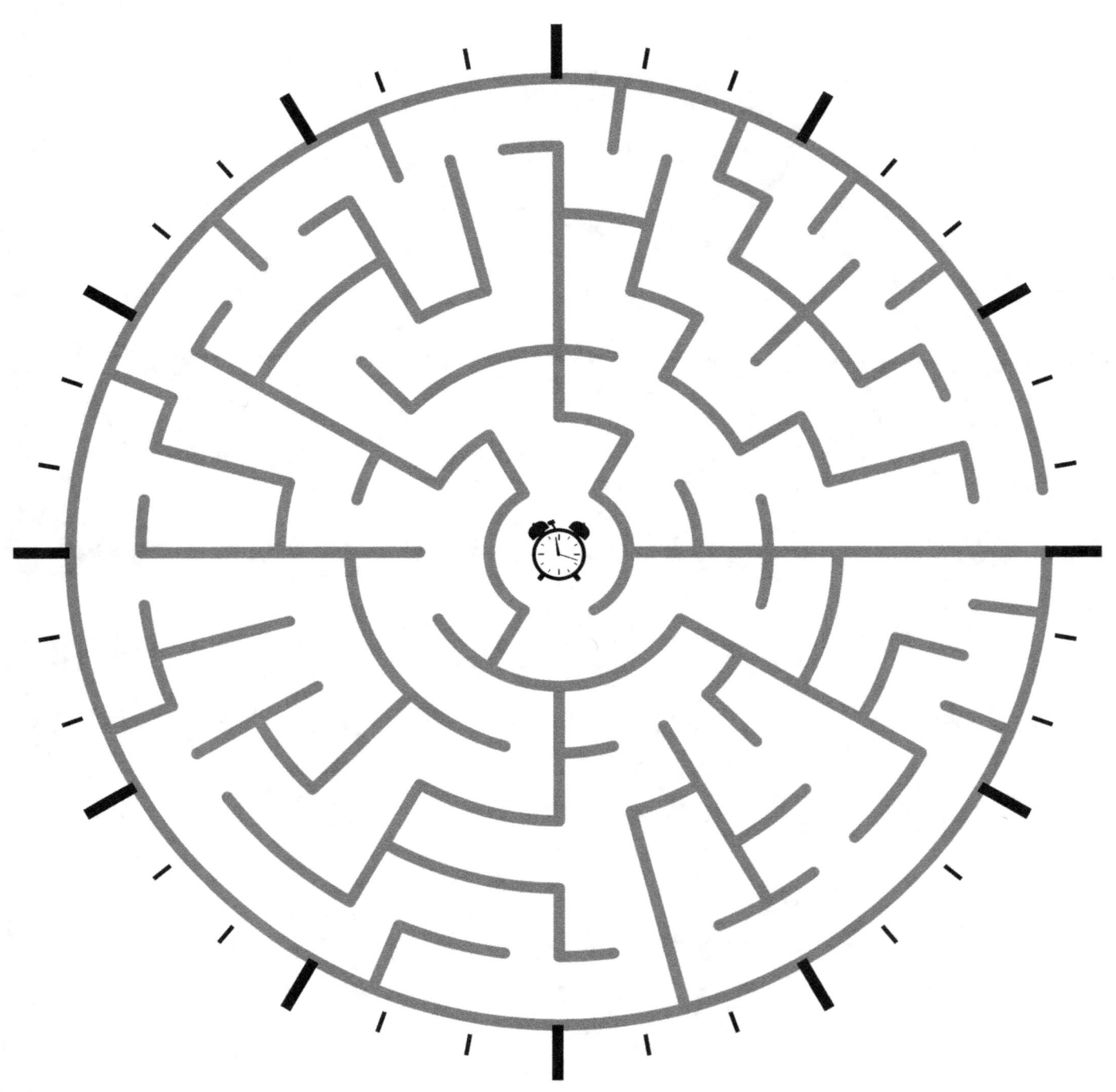

What Time is the Entrance:_____

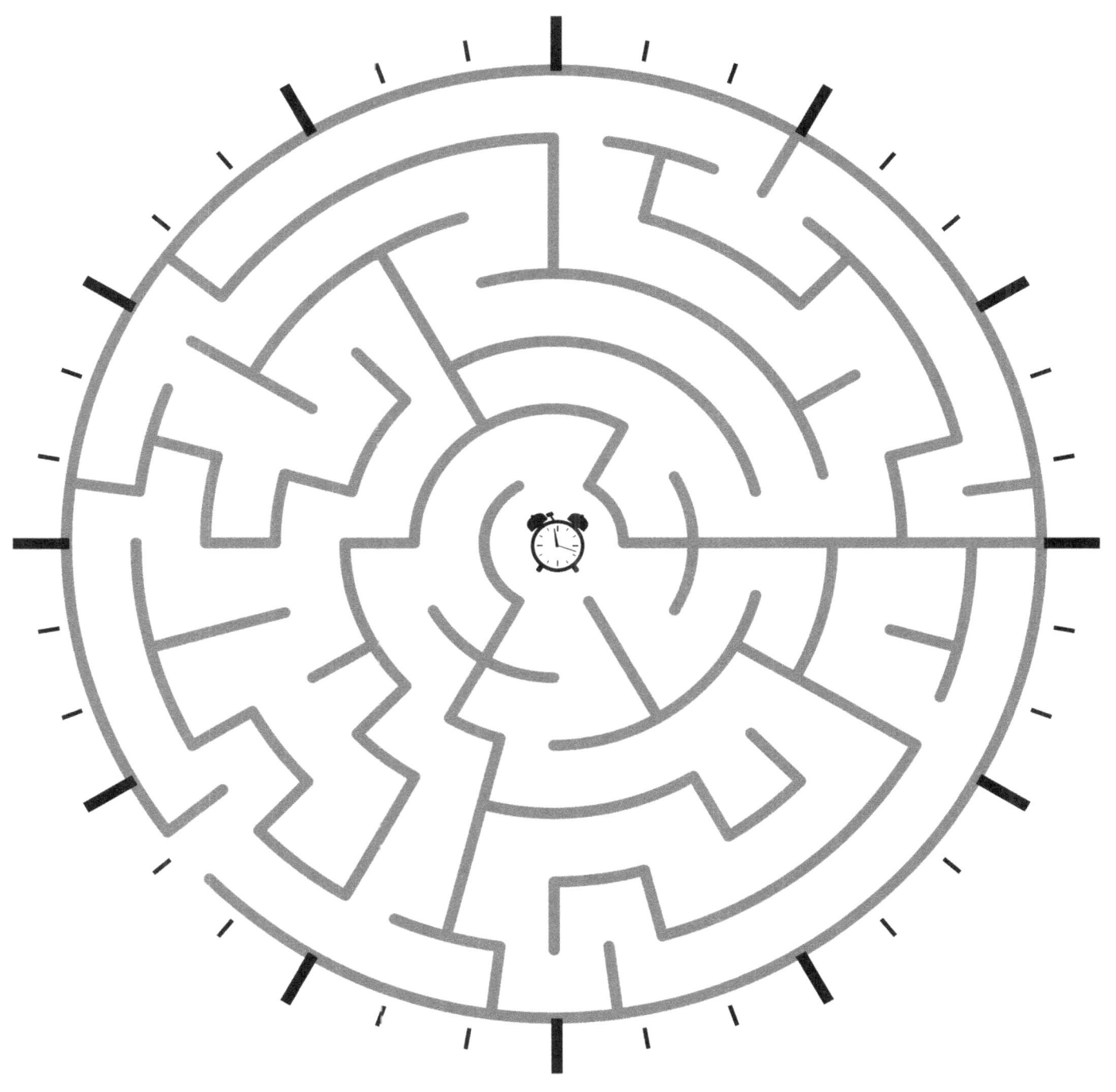

What Time is the Entrance:_____

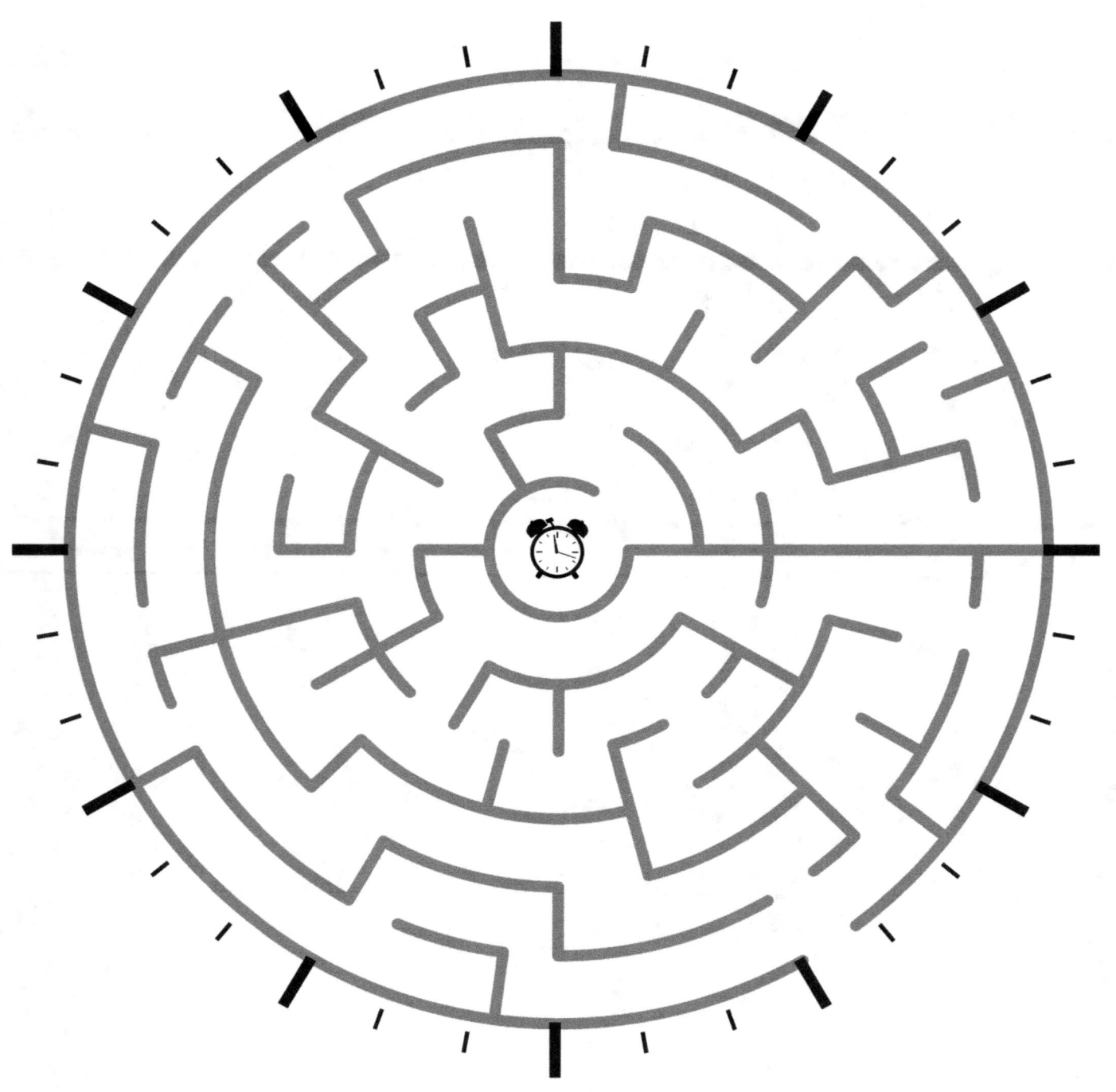

What Time is the Entrance:_____

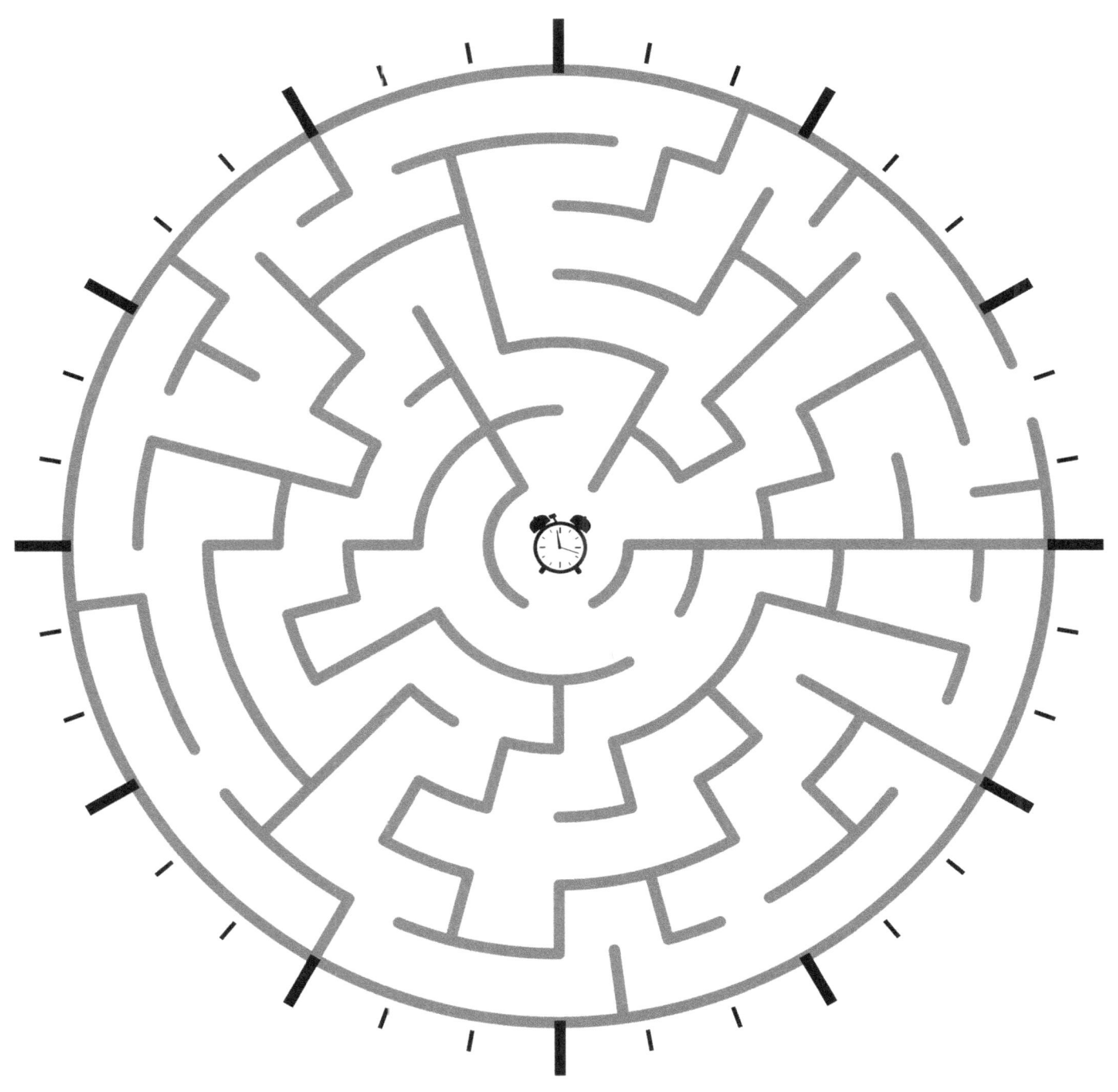

What Time is the Entrance: _____

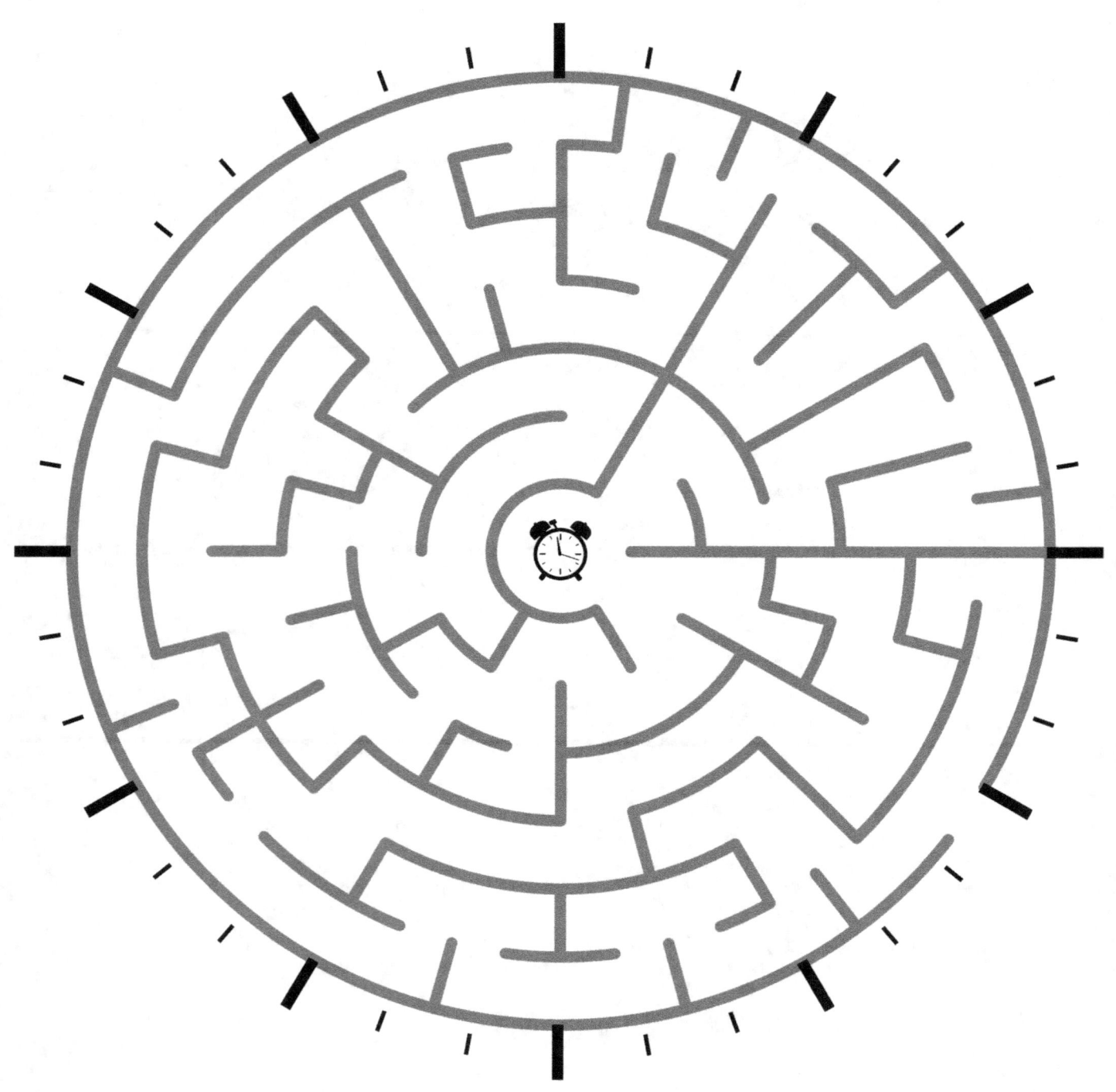

What Time is the Entrance:_____

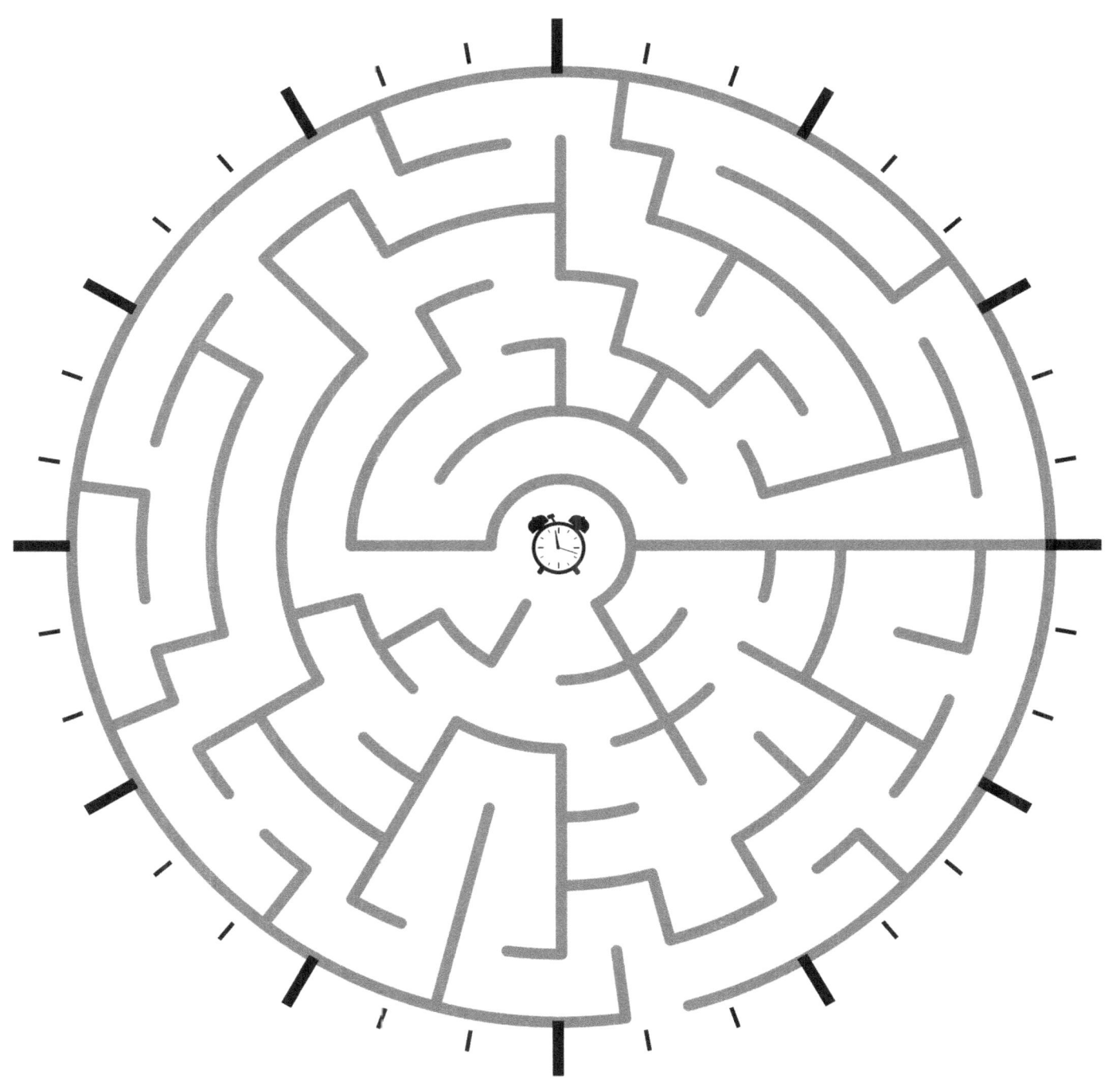

What Time is the Entrance:_____

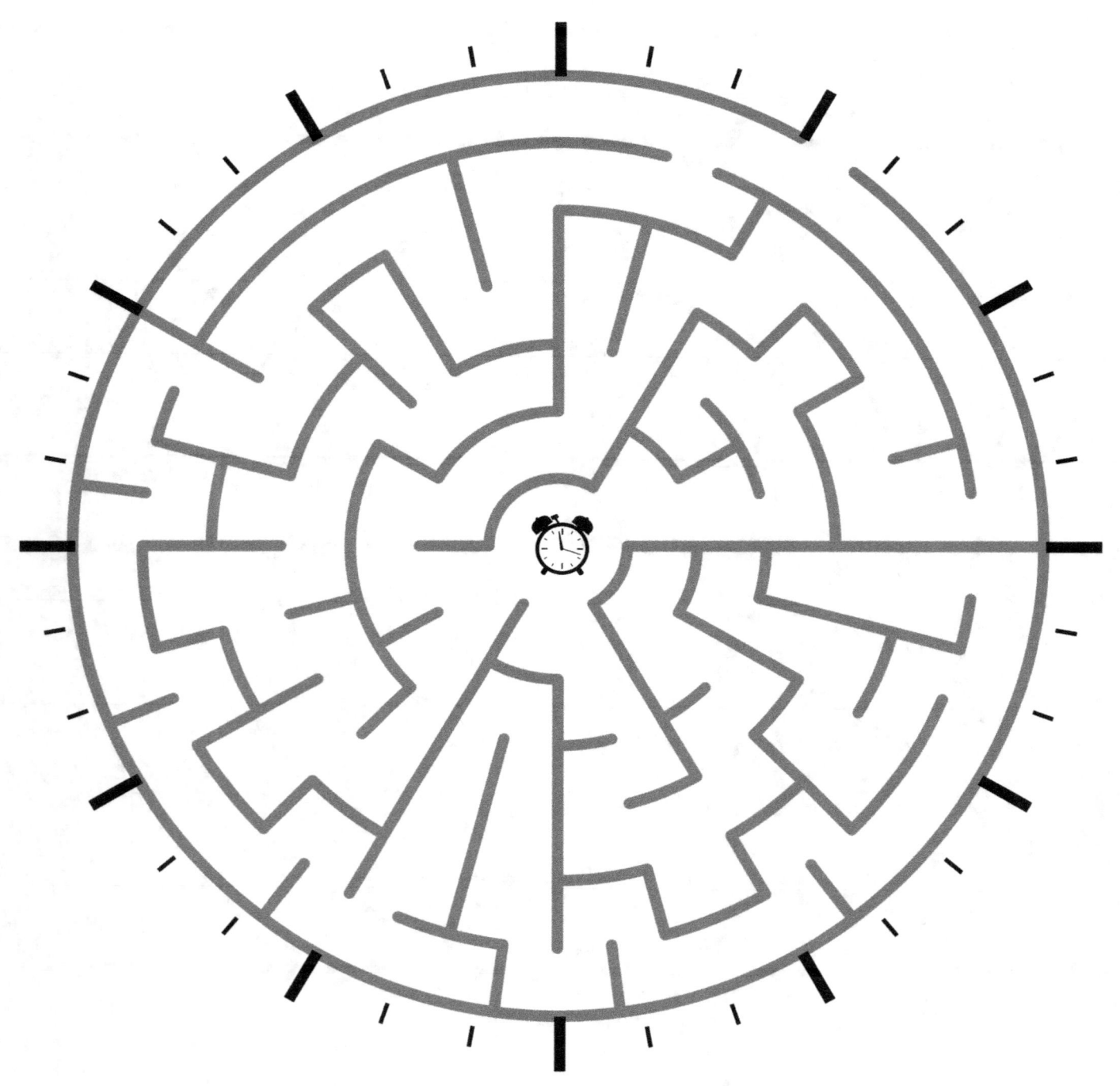

What Time is the Entrance:_____

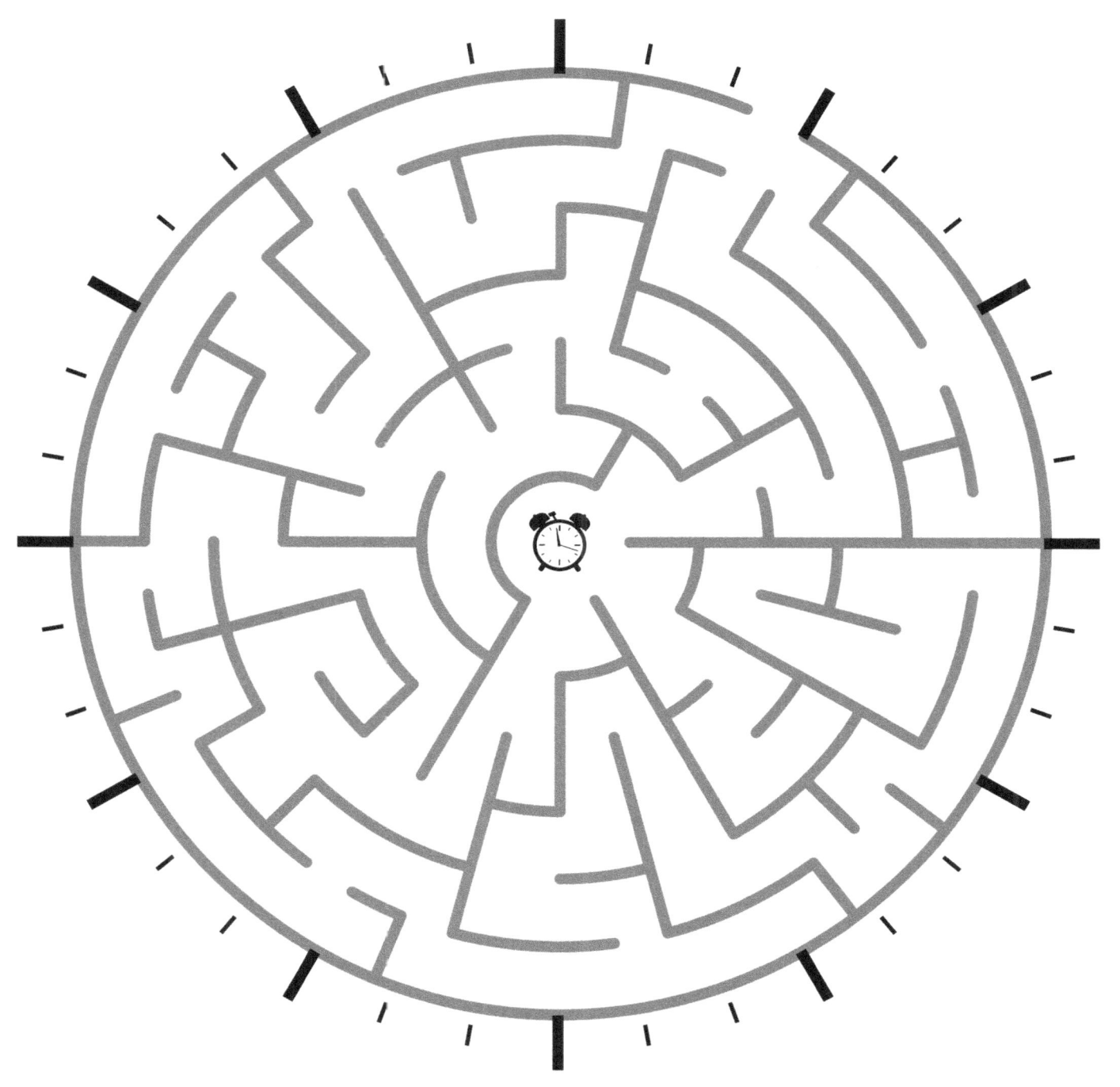

What Time is the Entrance:_____

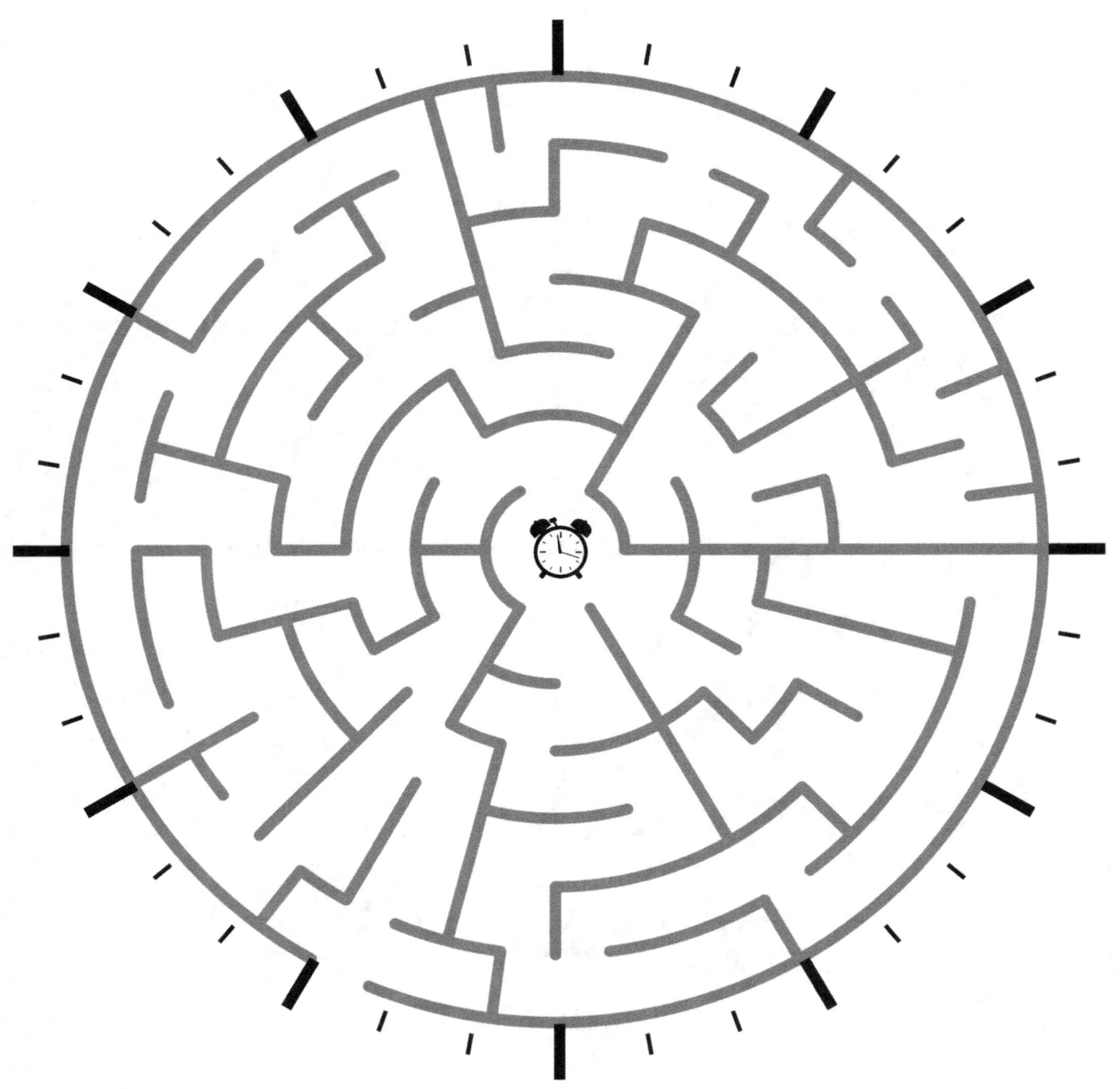

What Time is the Entrance:_____

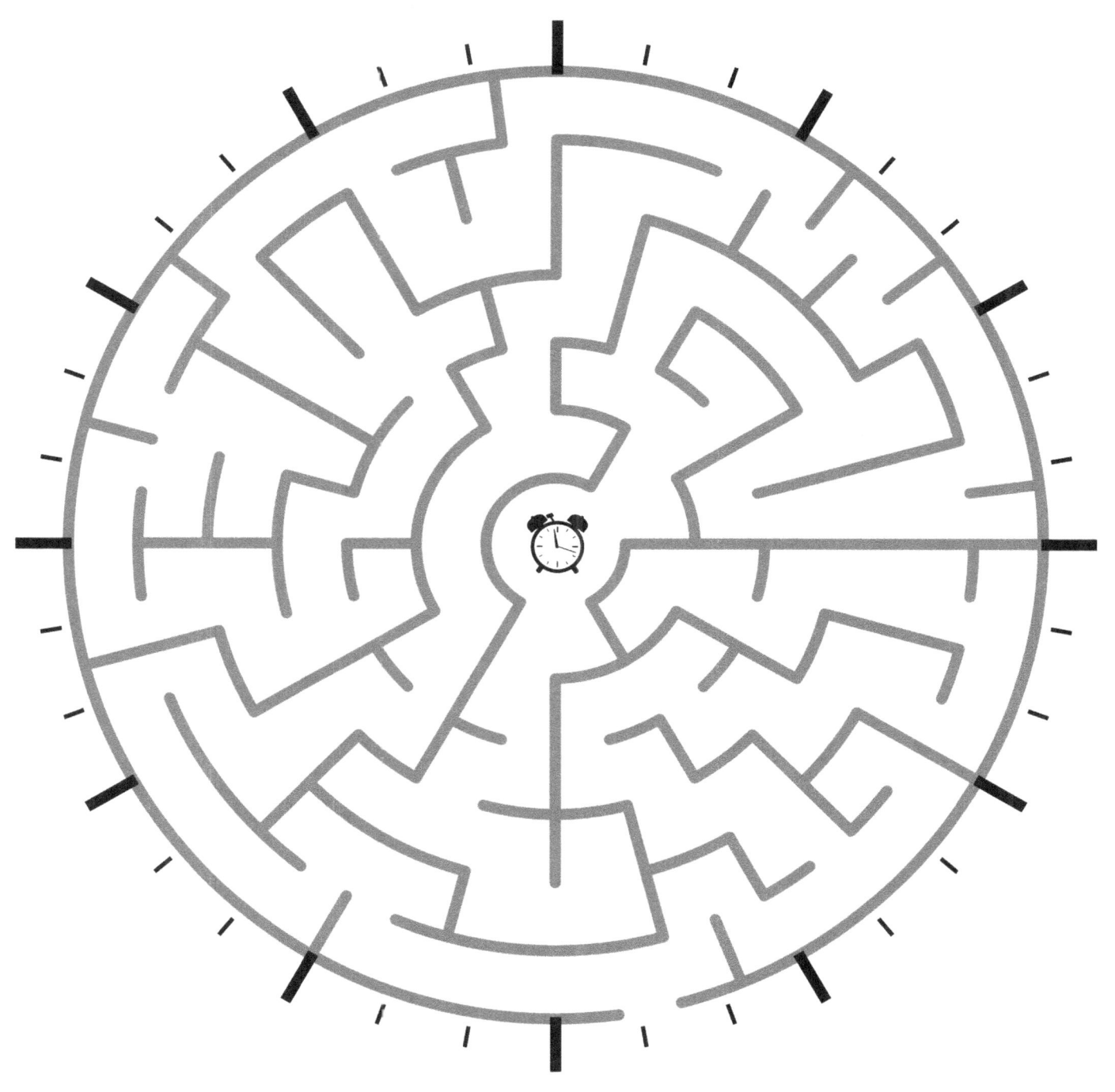

What Time is the Entrance:_____

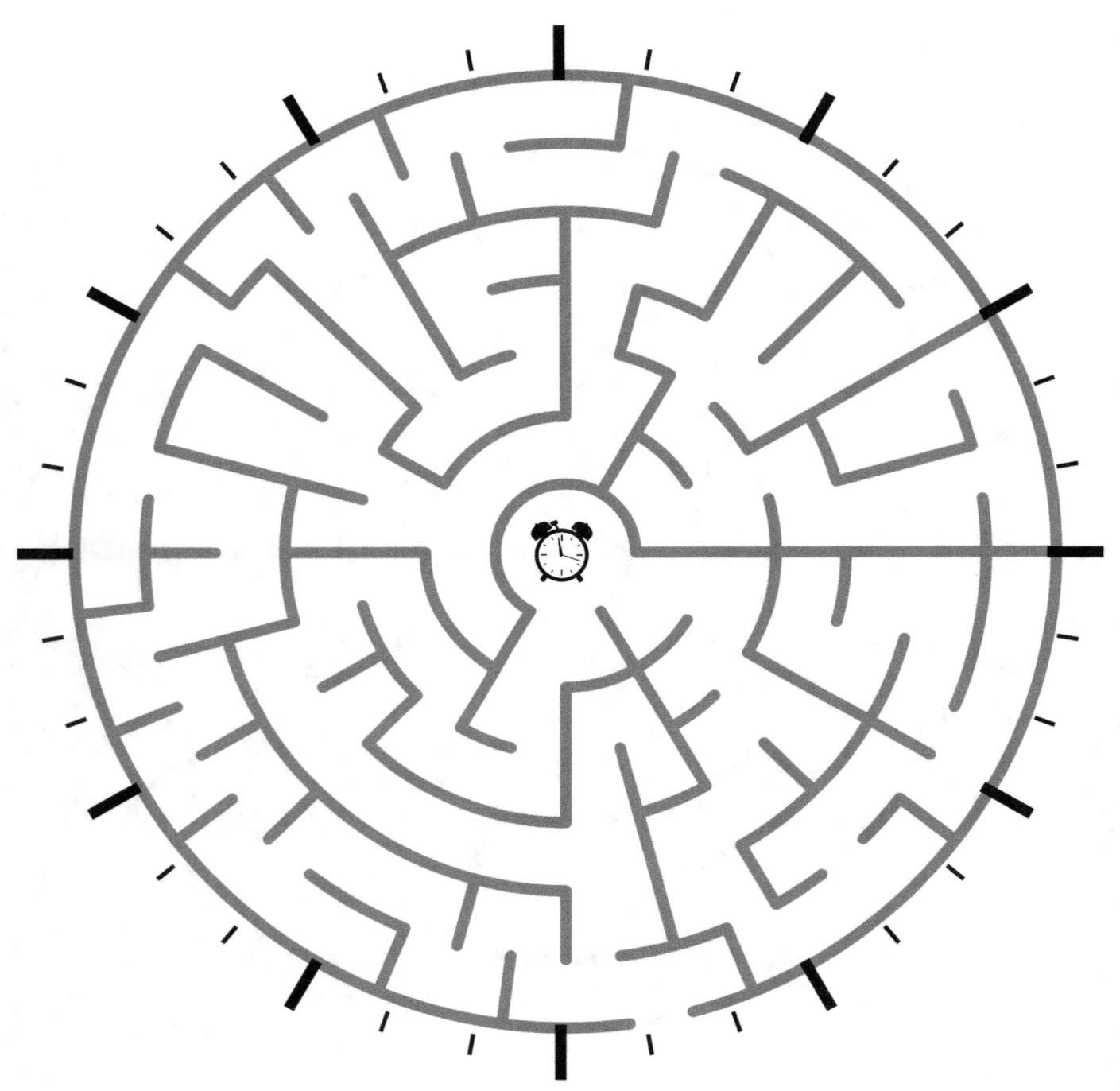

What Time is the Entrance:_____

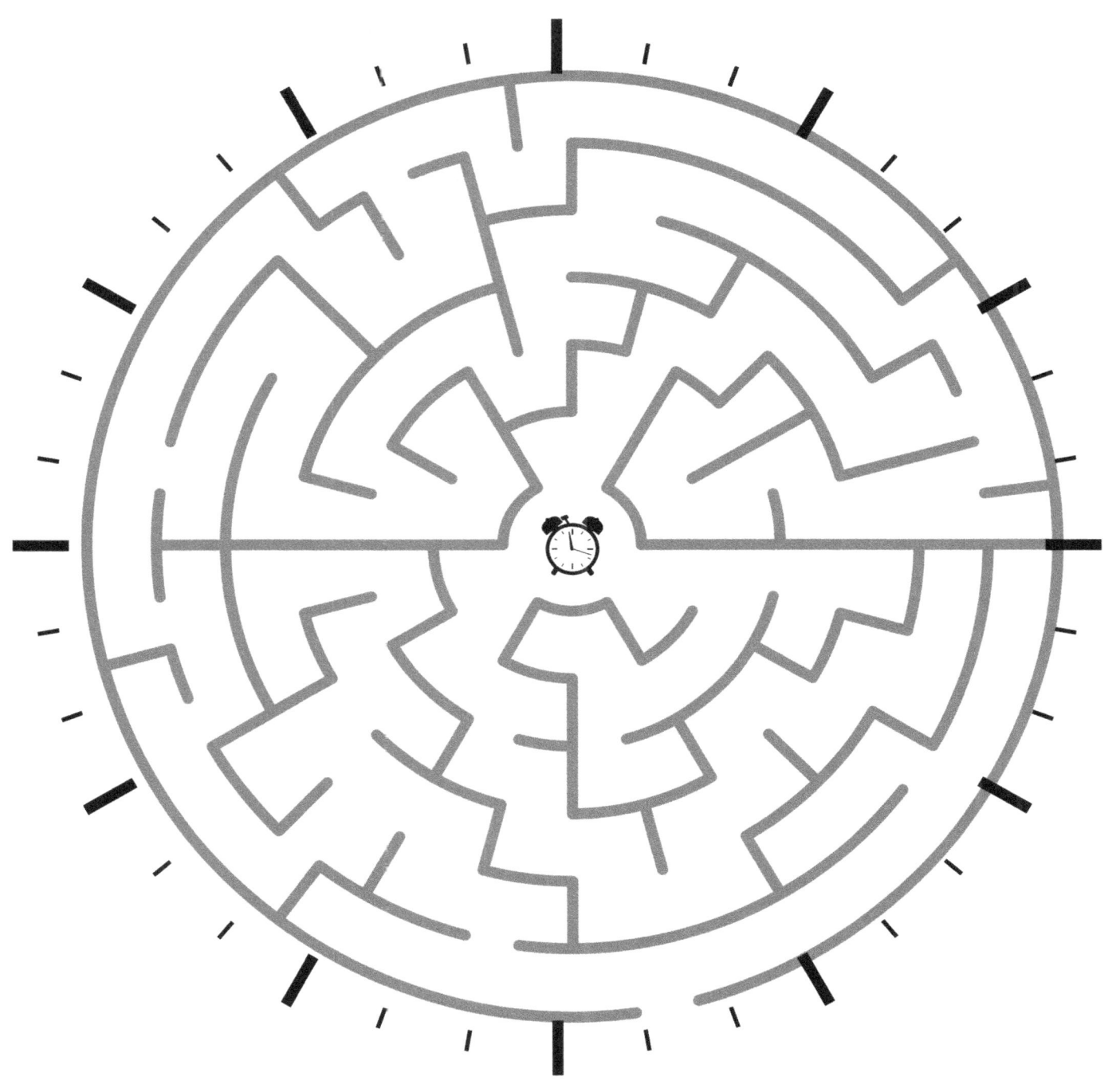

What Time is the Entrance:_____

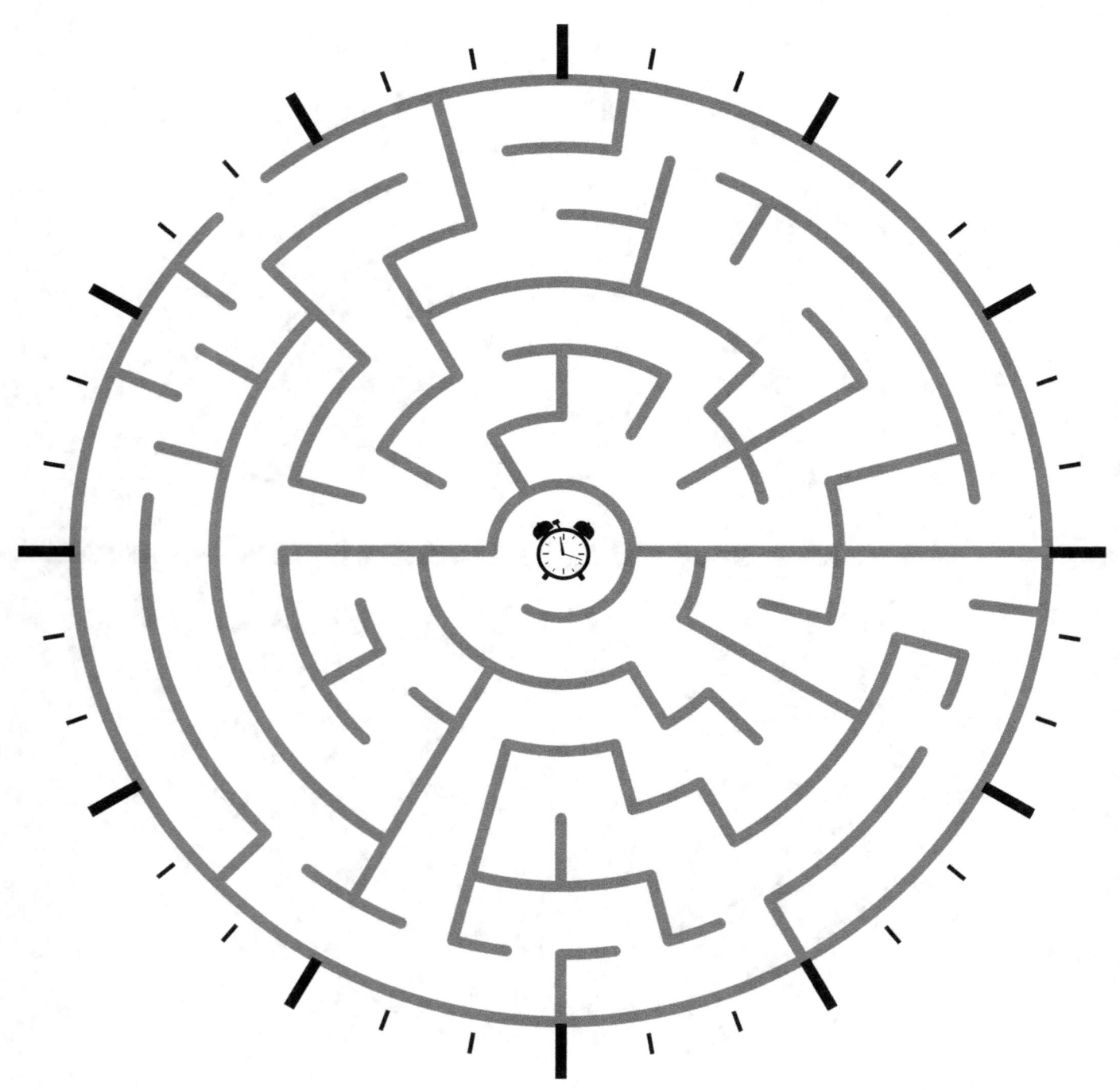

What Time is the Entrance:_____

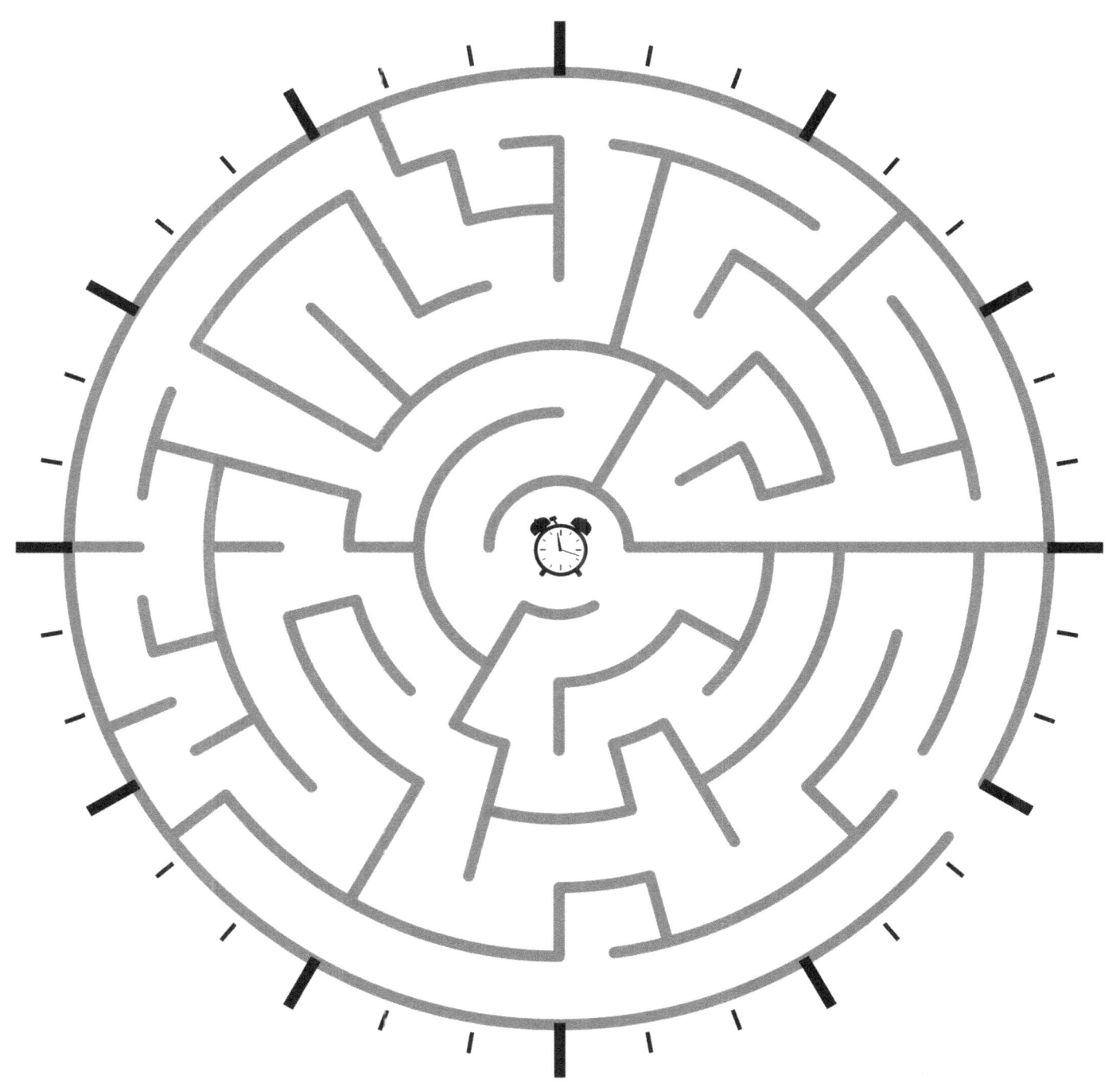

What Time is the Entrance:_____

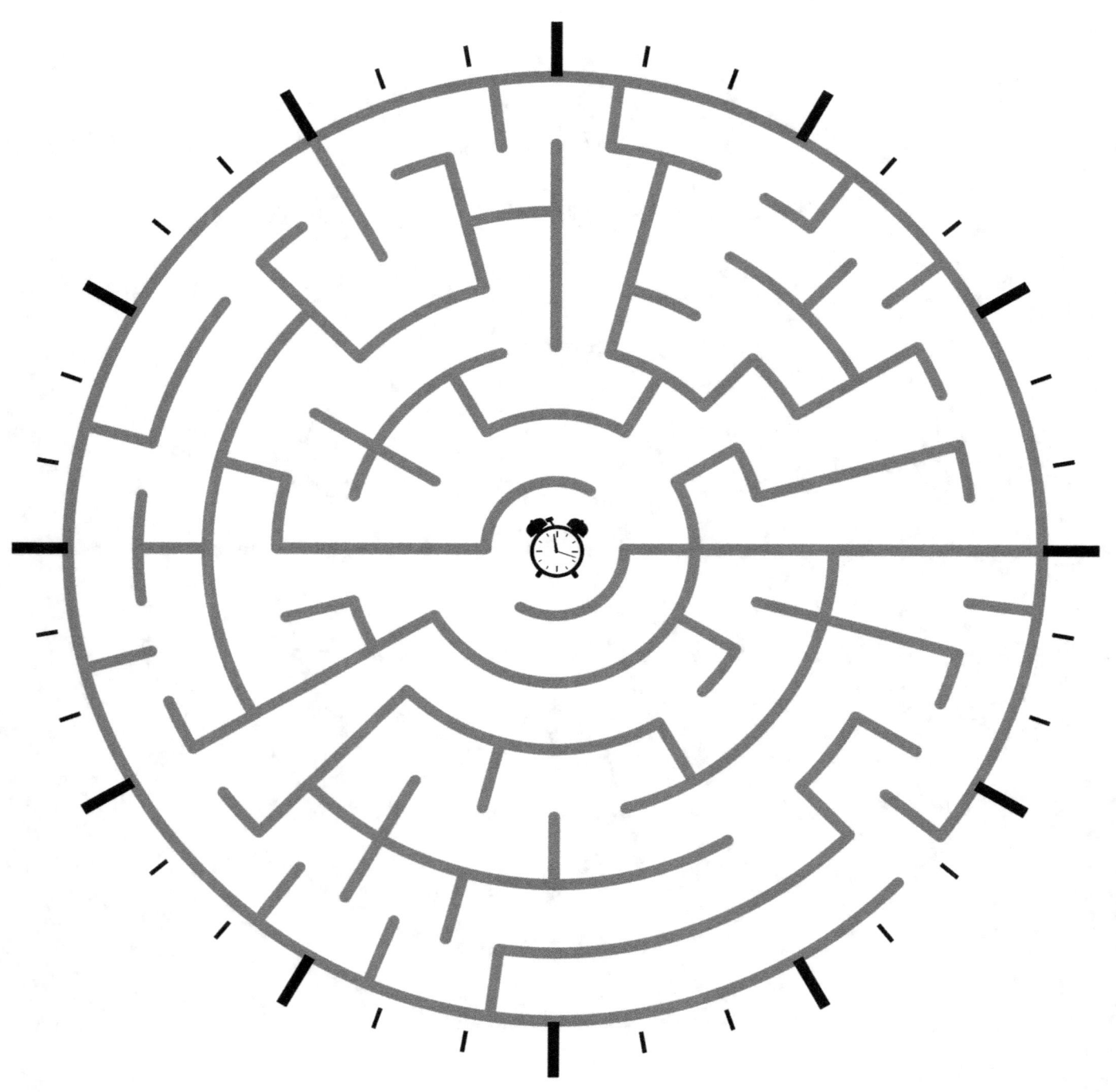

What Time is the Entrance:_____

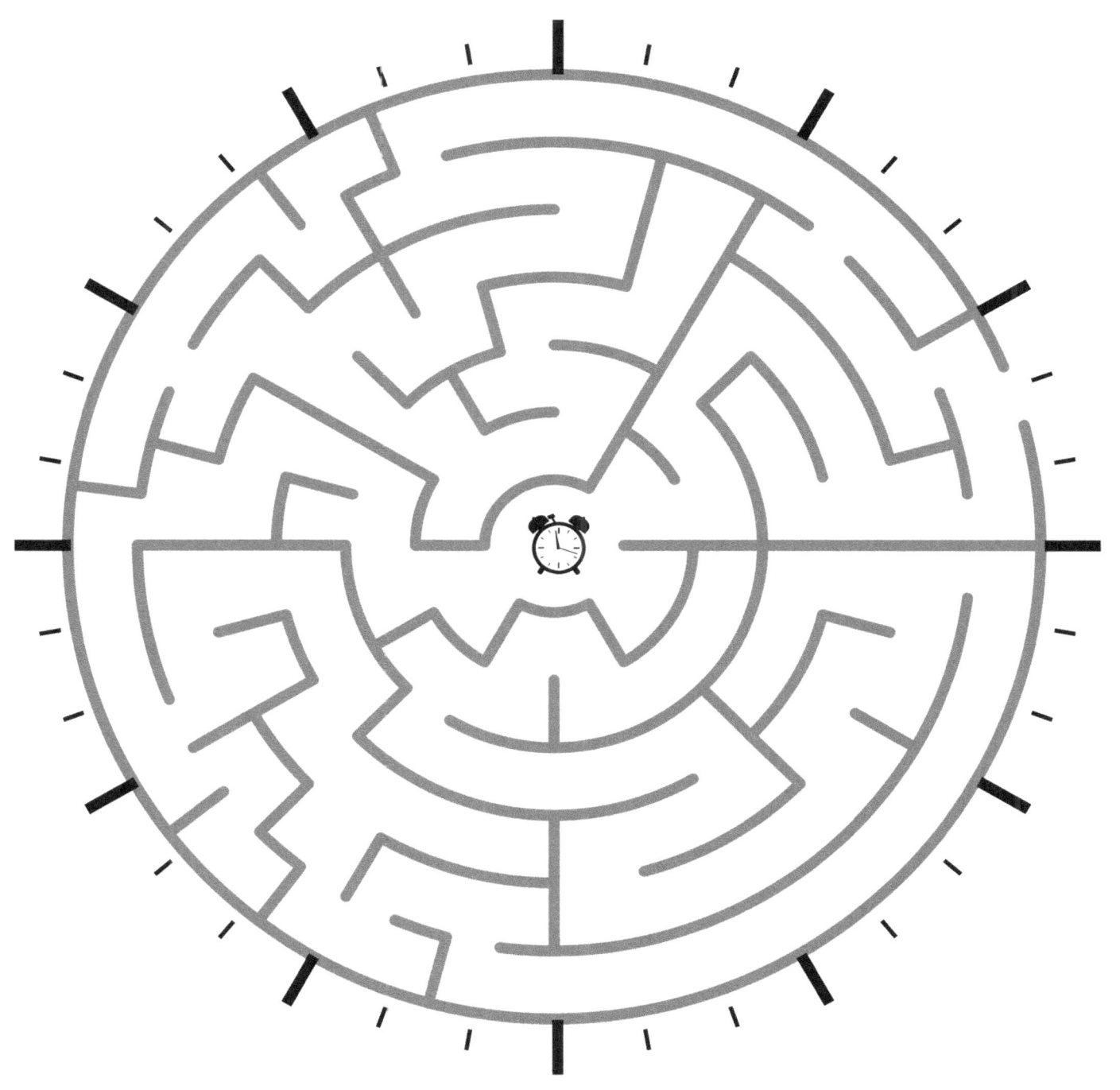

What Time is the Entrance:_____

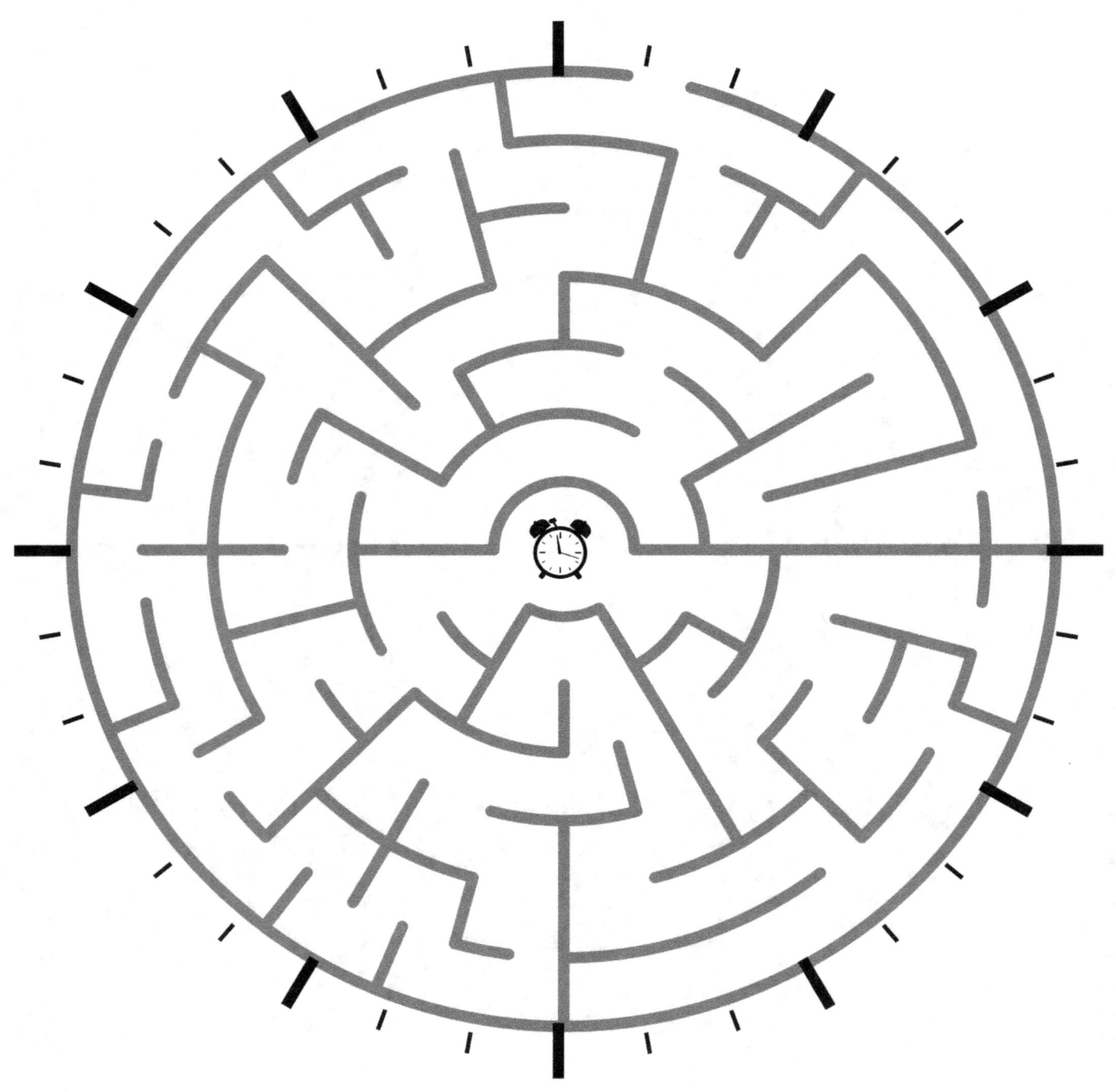

What Time is the Entrance:_____

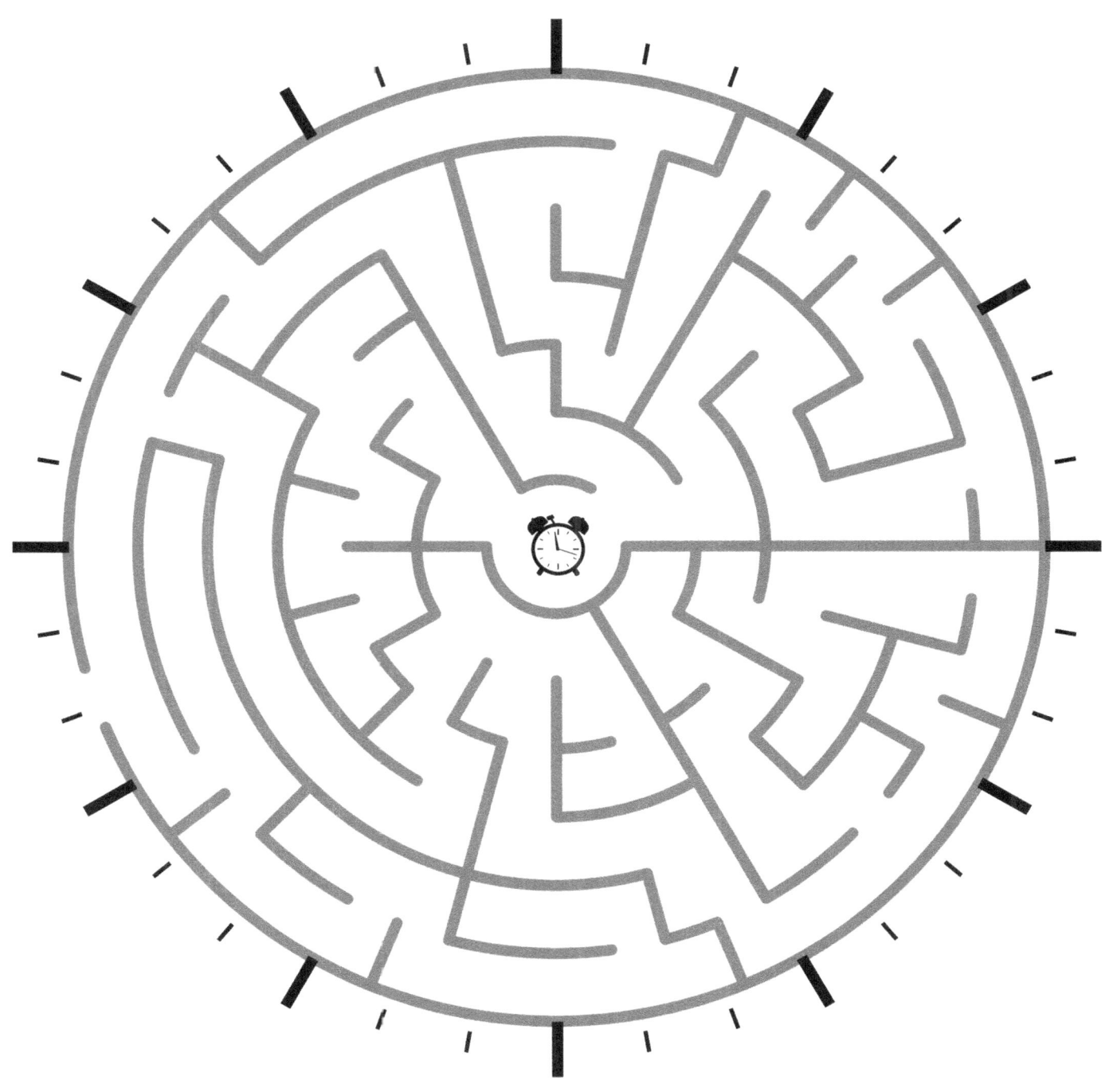

What Time is the Entrance:_____

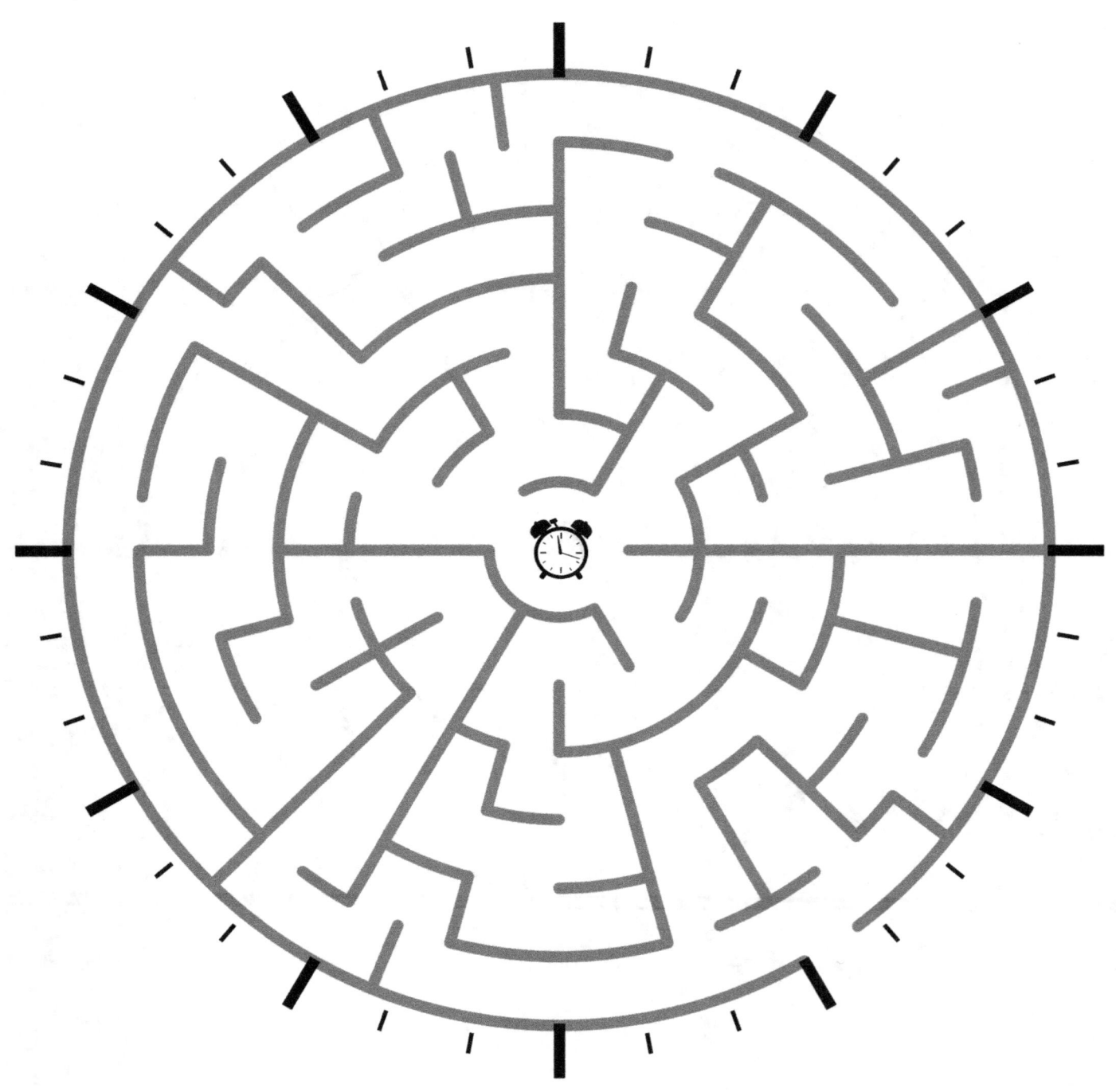

What Time is the Entrance:_____

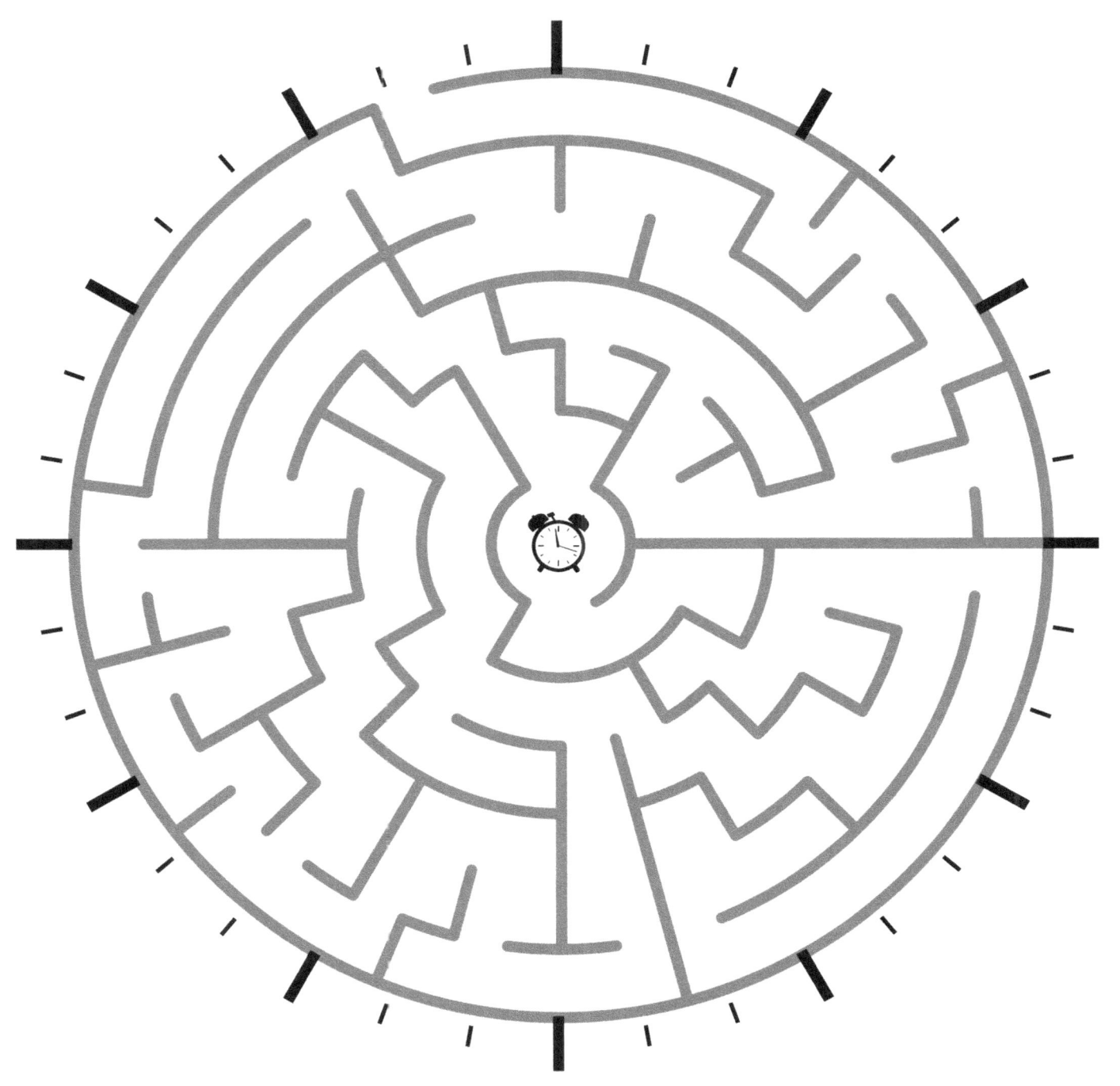

What Time is the Entrance:_____

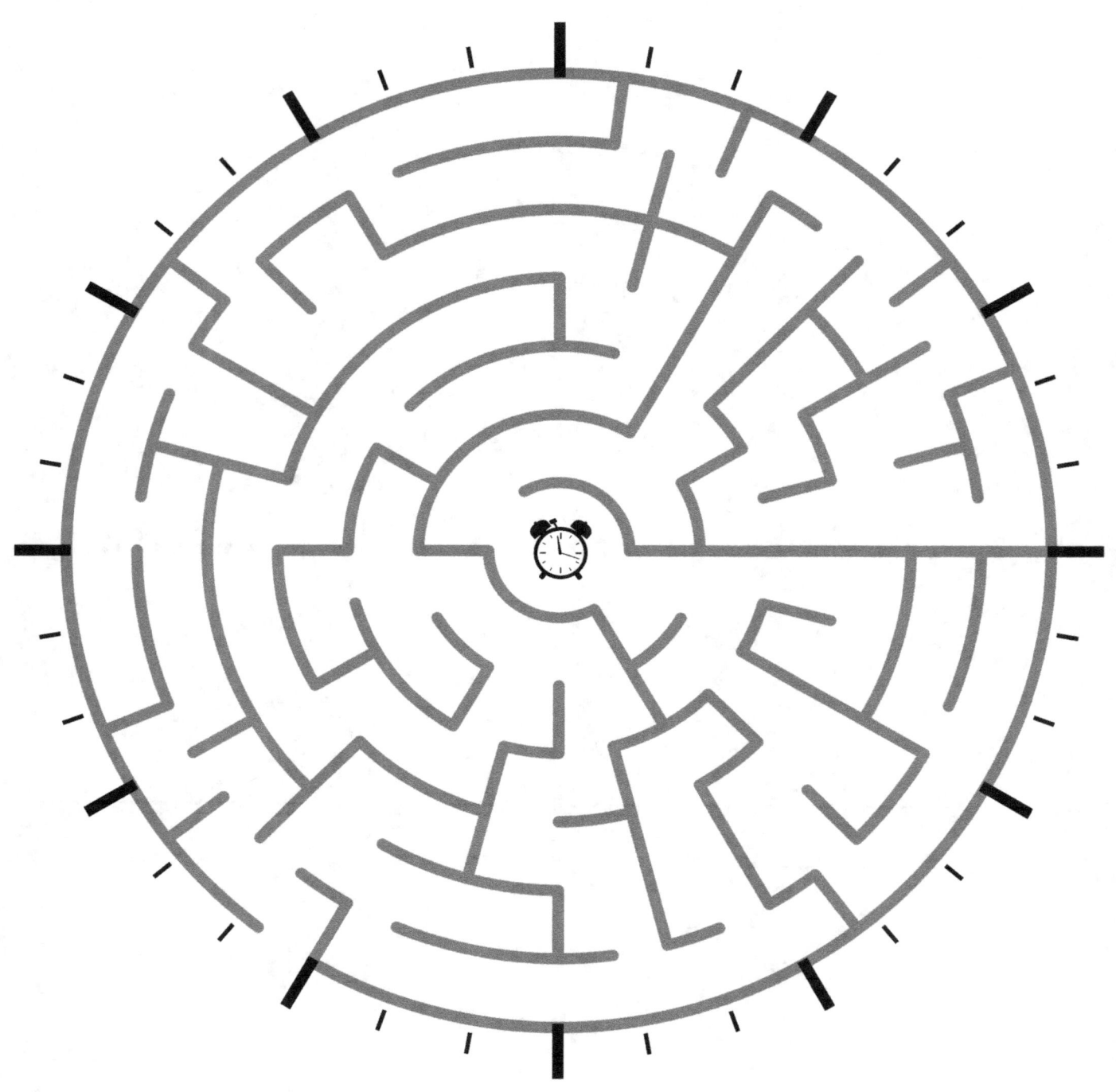

What Time is the Entrance:_____

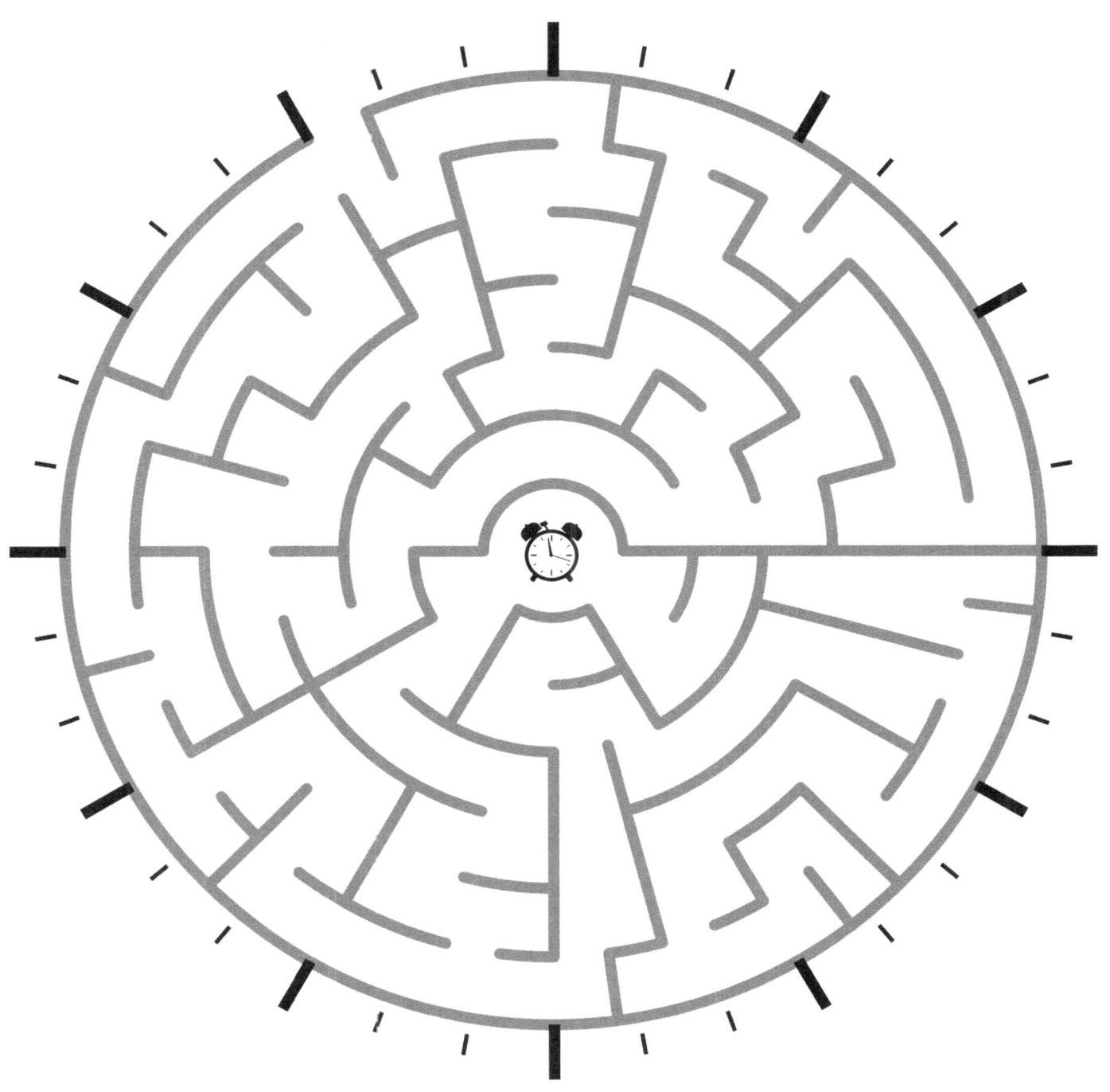

What Time is the Entrance:_____

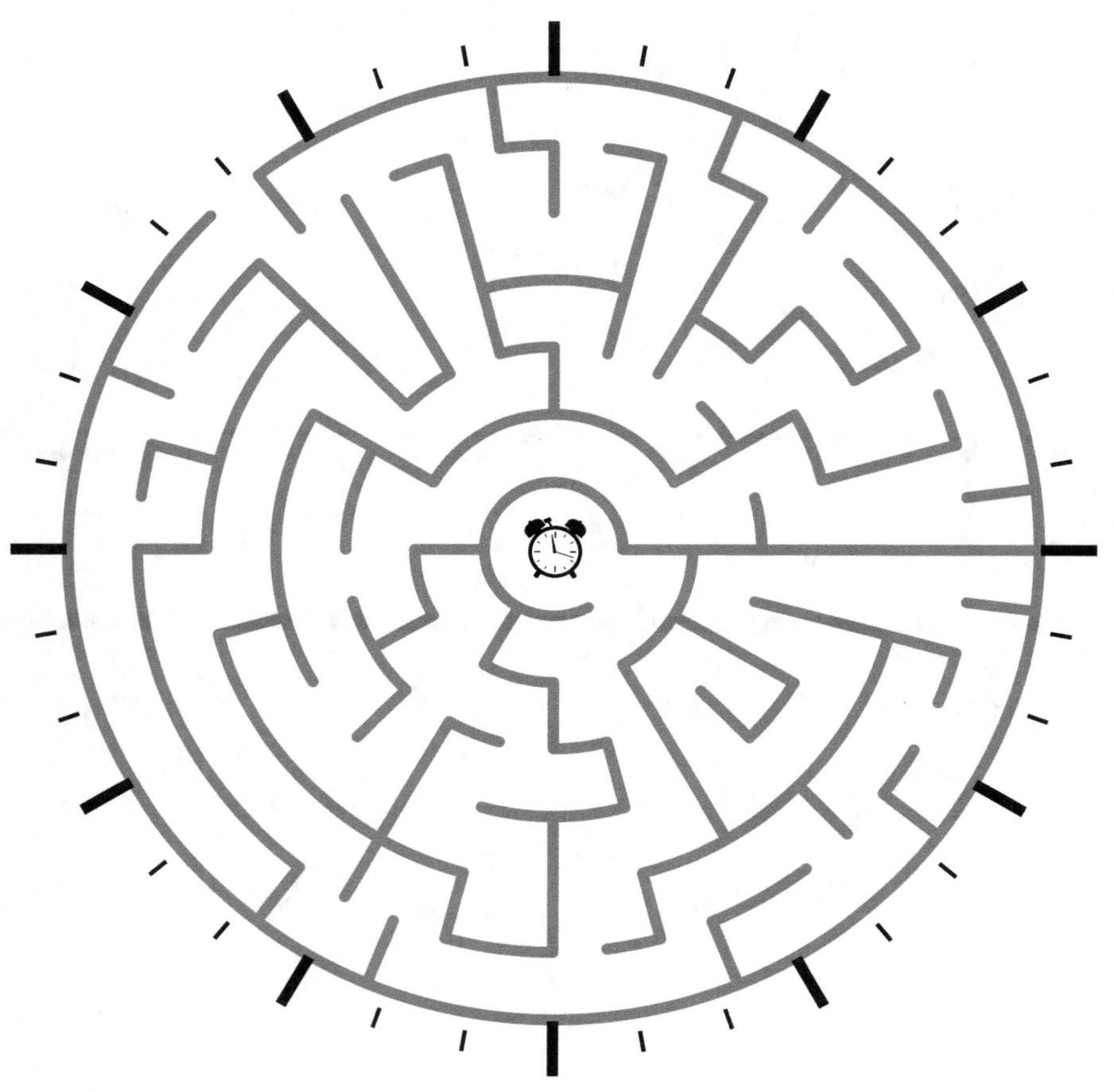

What Time is the Entrance:_____

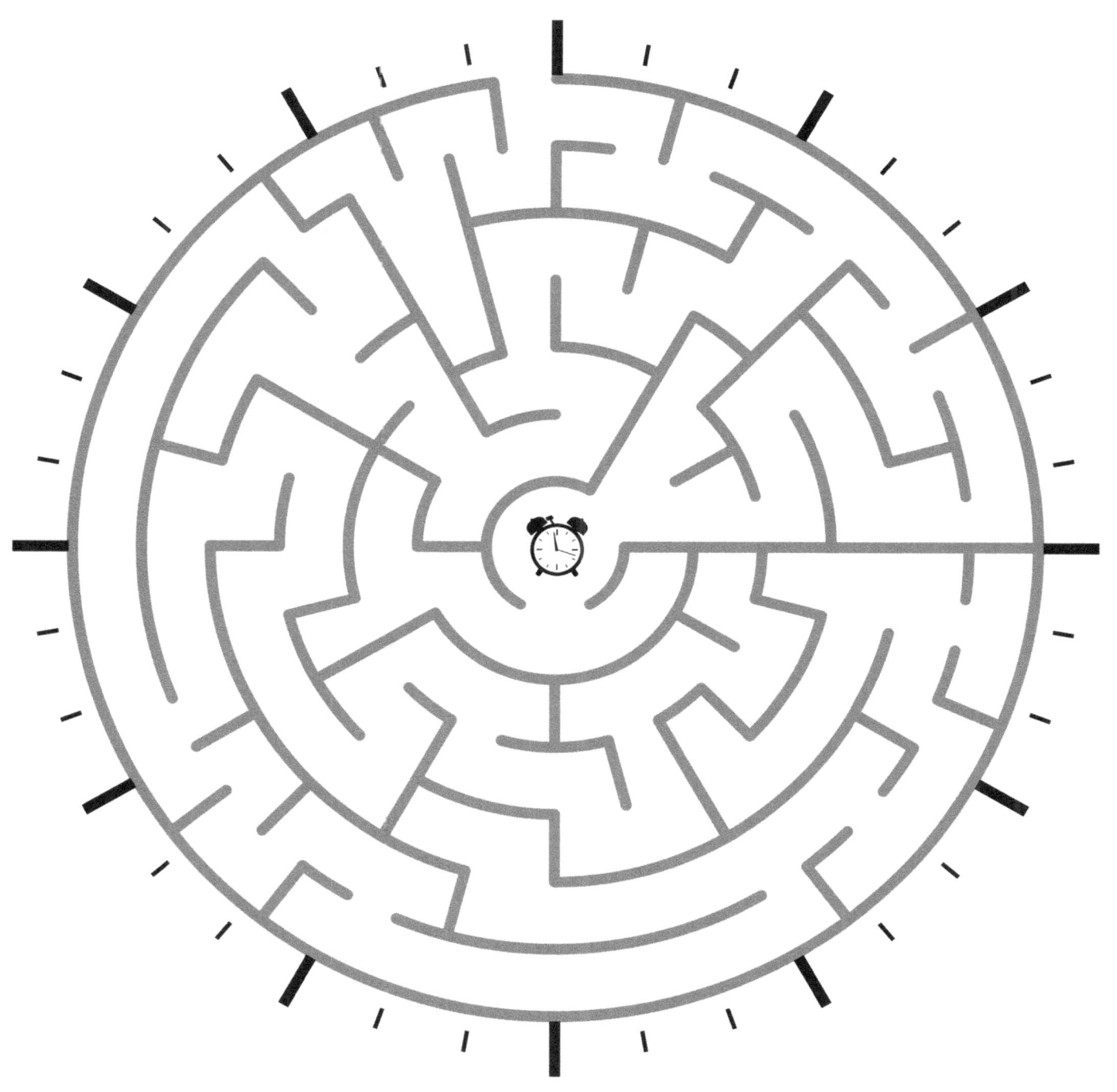

What Time is the Entrance:_____

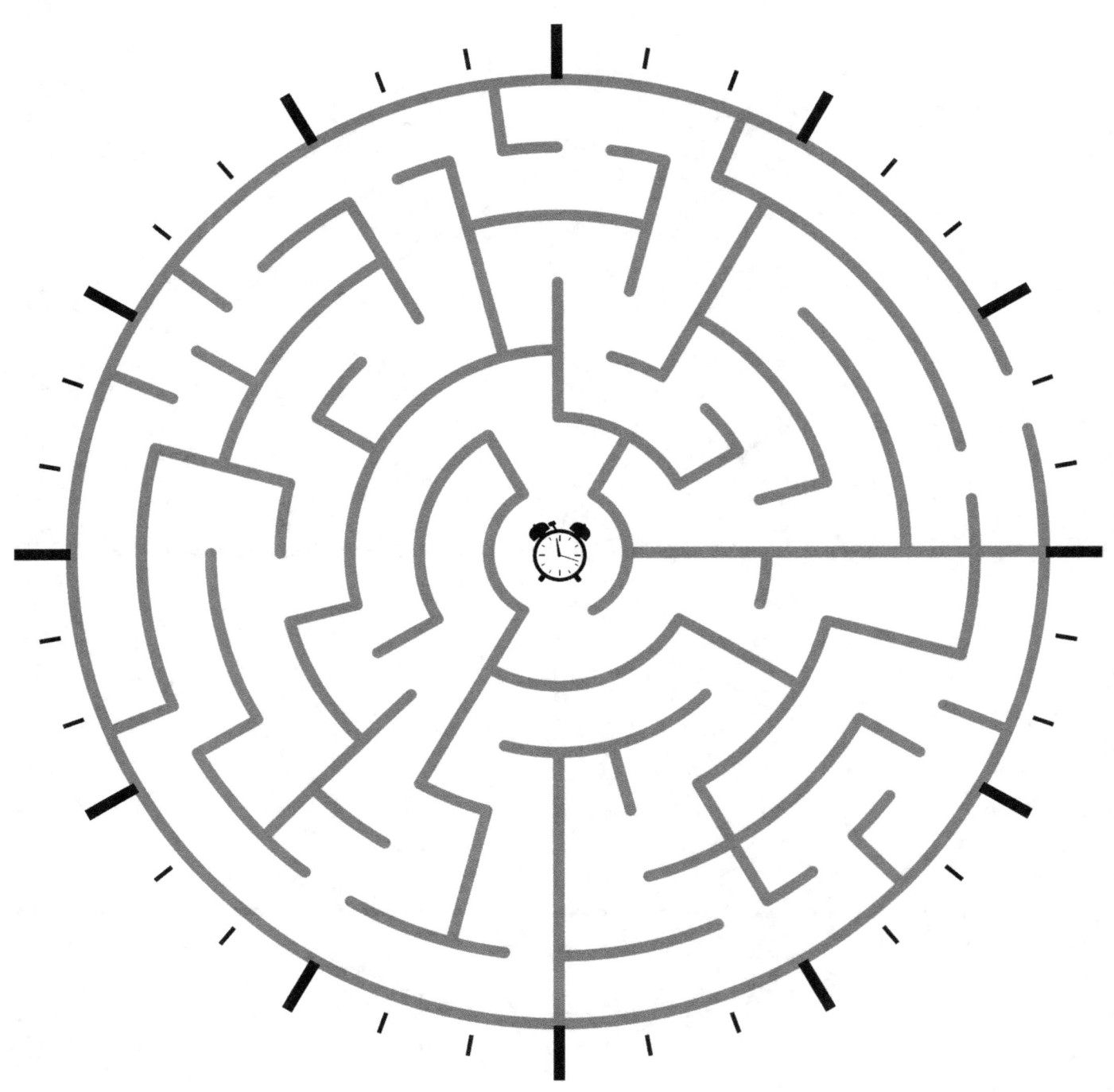

What Time is the Entrance:_____

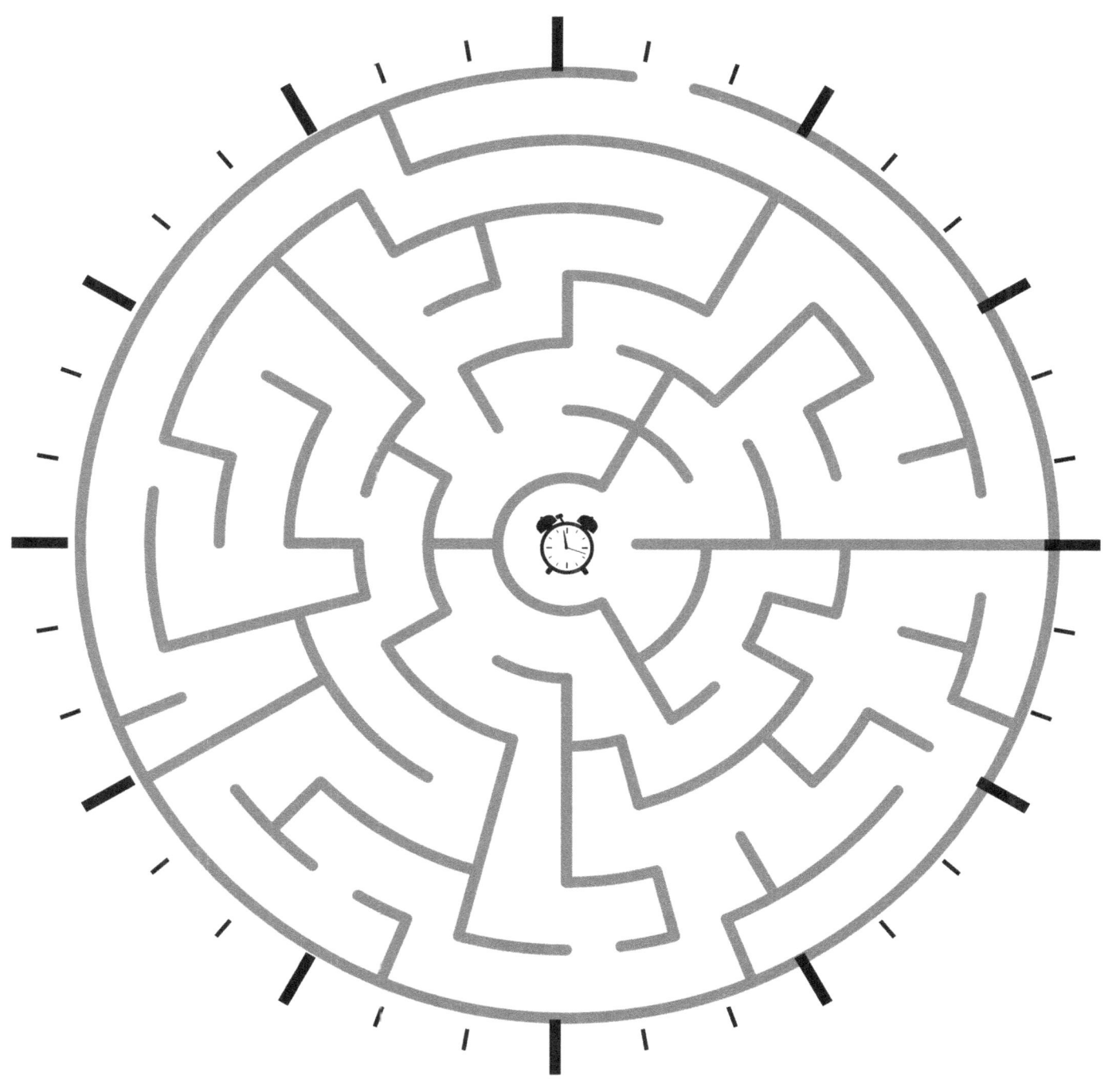

What Time is the Entrance:_____

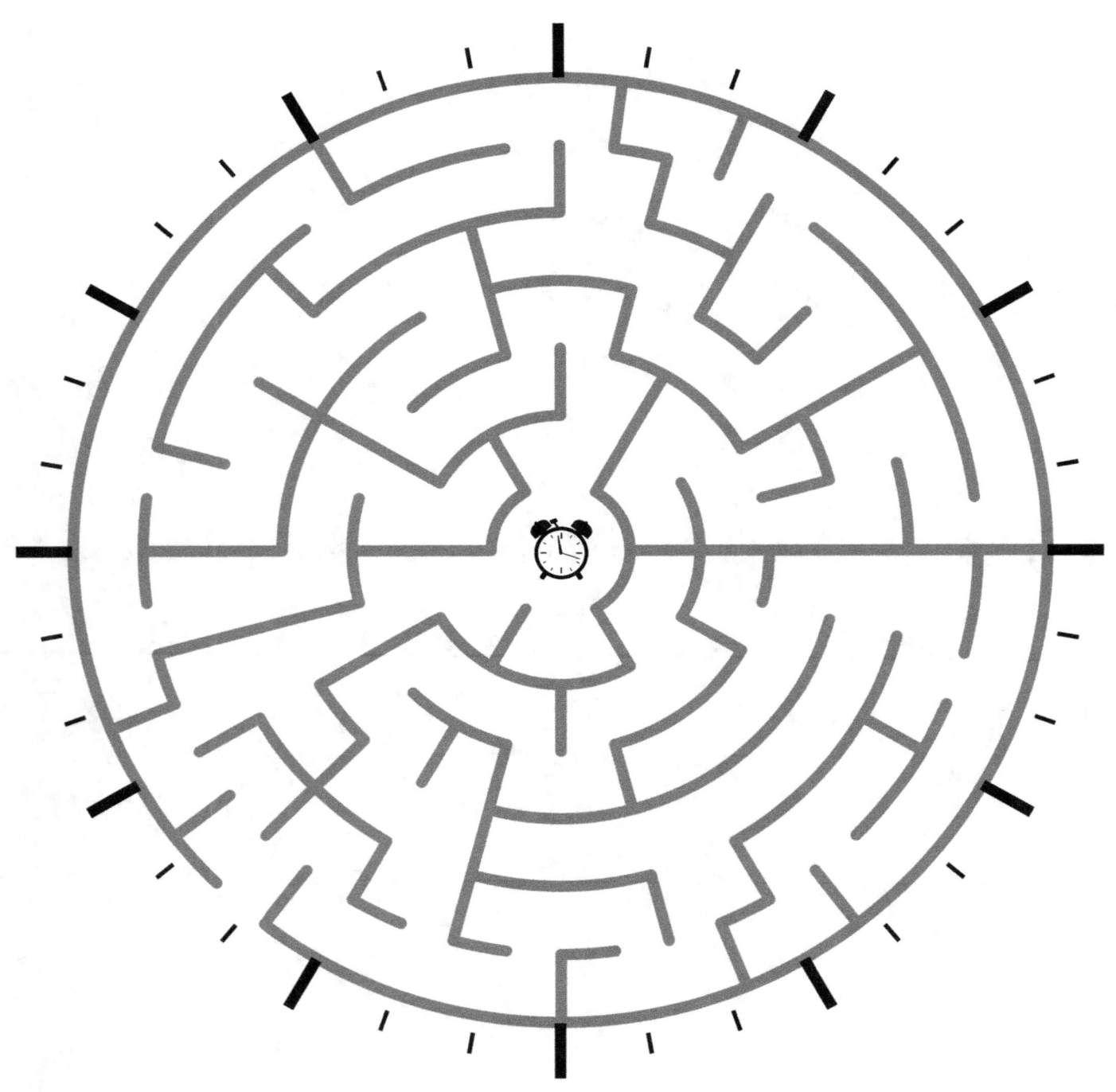

What Time is the Entrance:_____

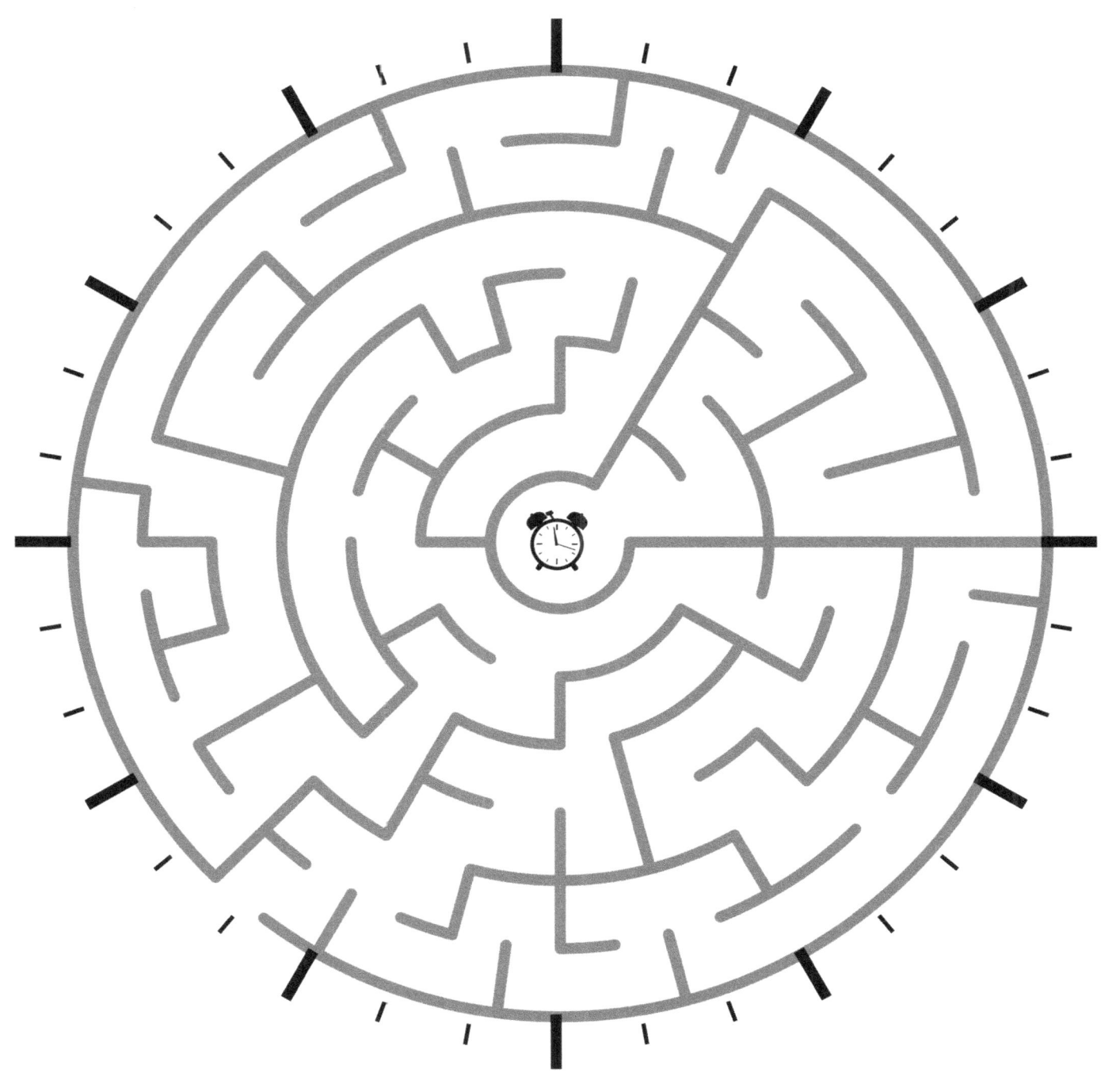

What Time is the Entrance:_____

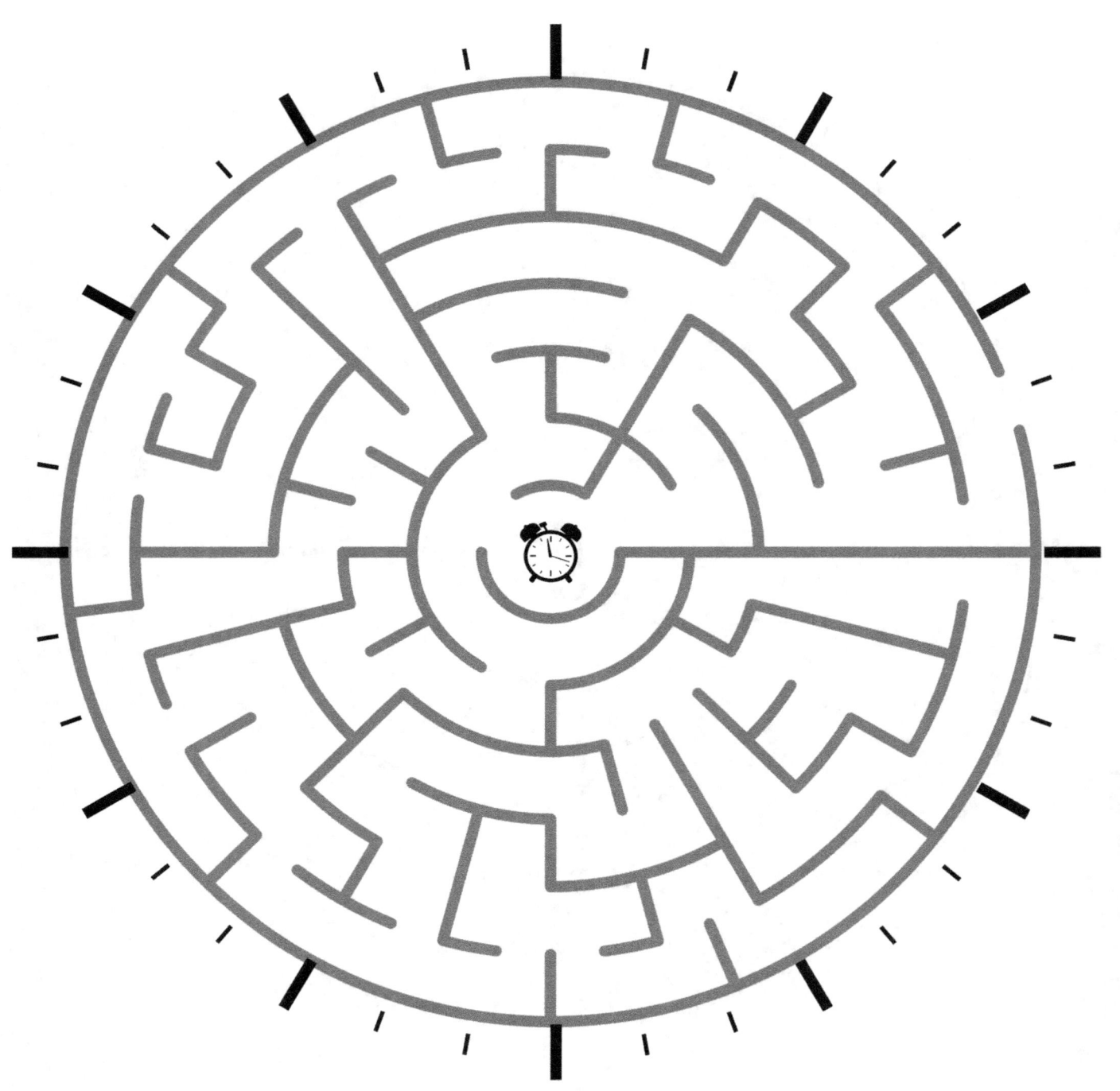

What Time is the Entrance:_____

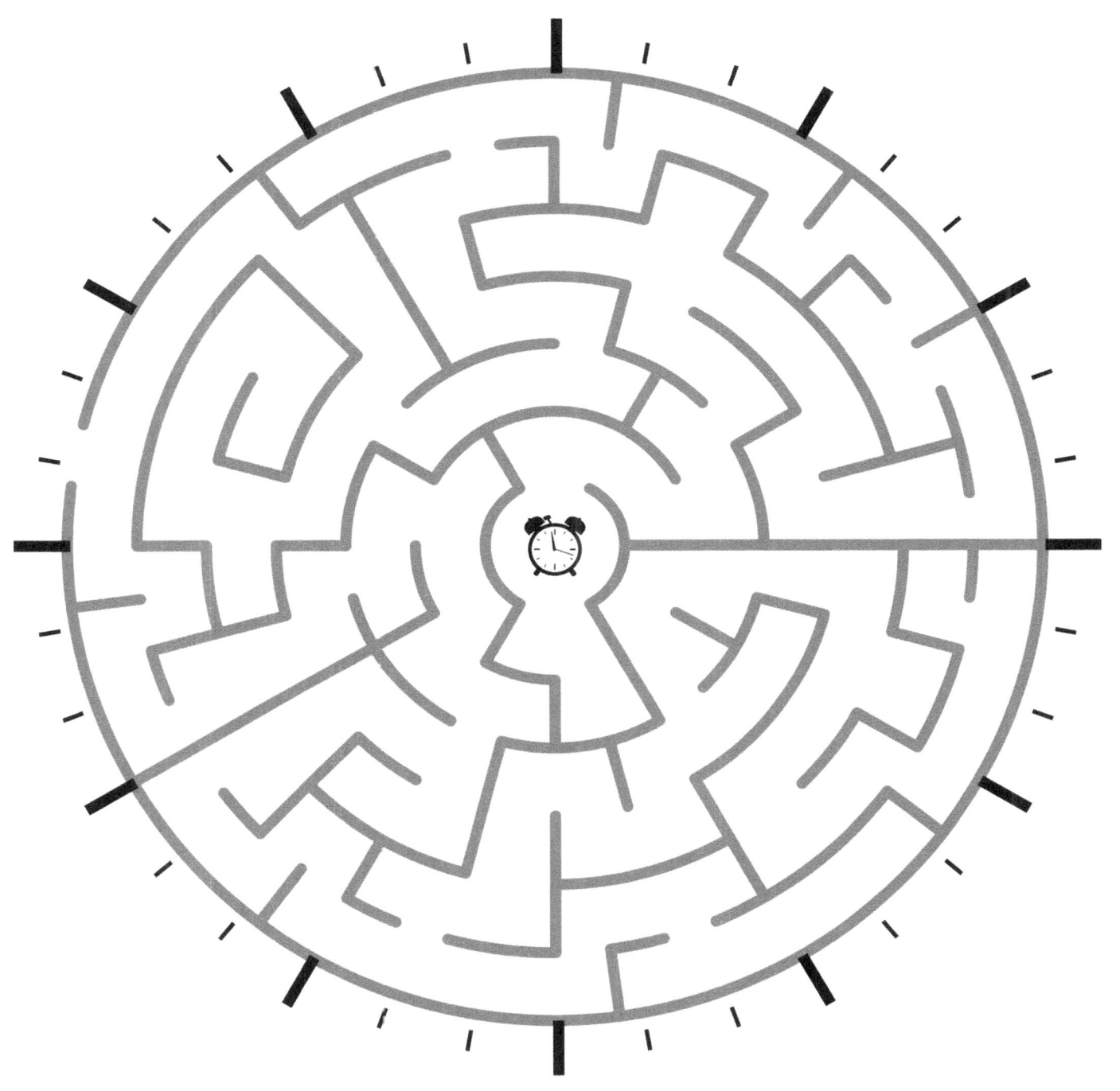

What Time is the Entrance:_____

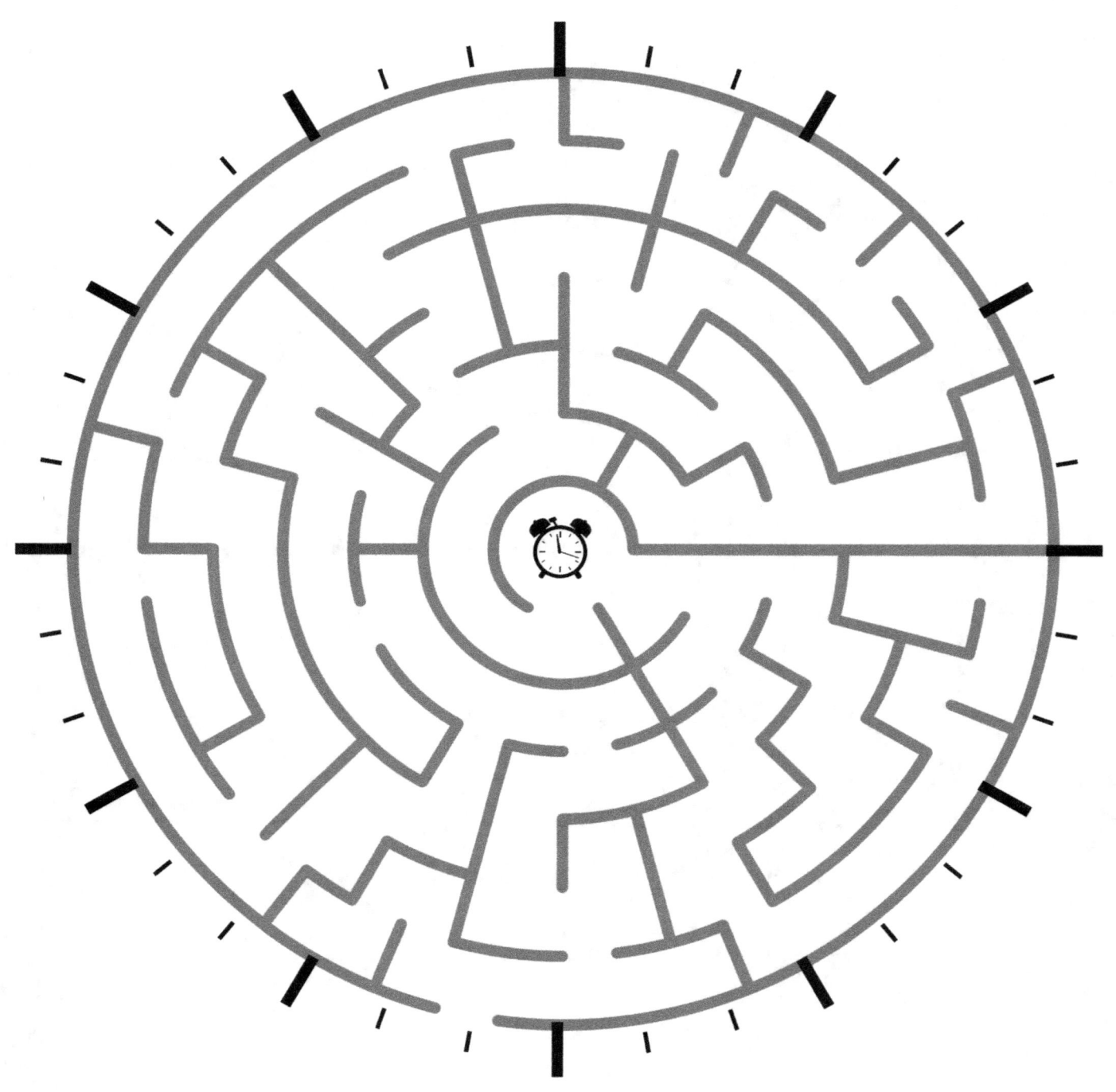

What Time is the Entrance:_____

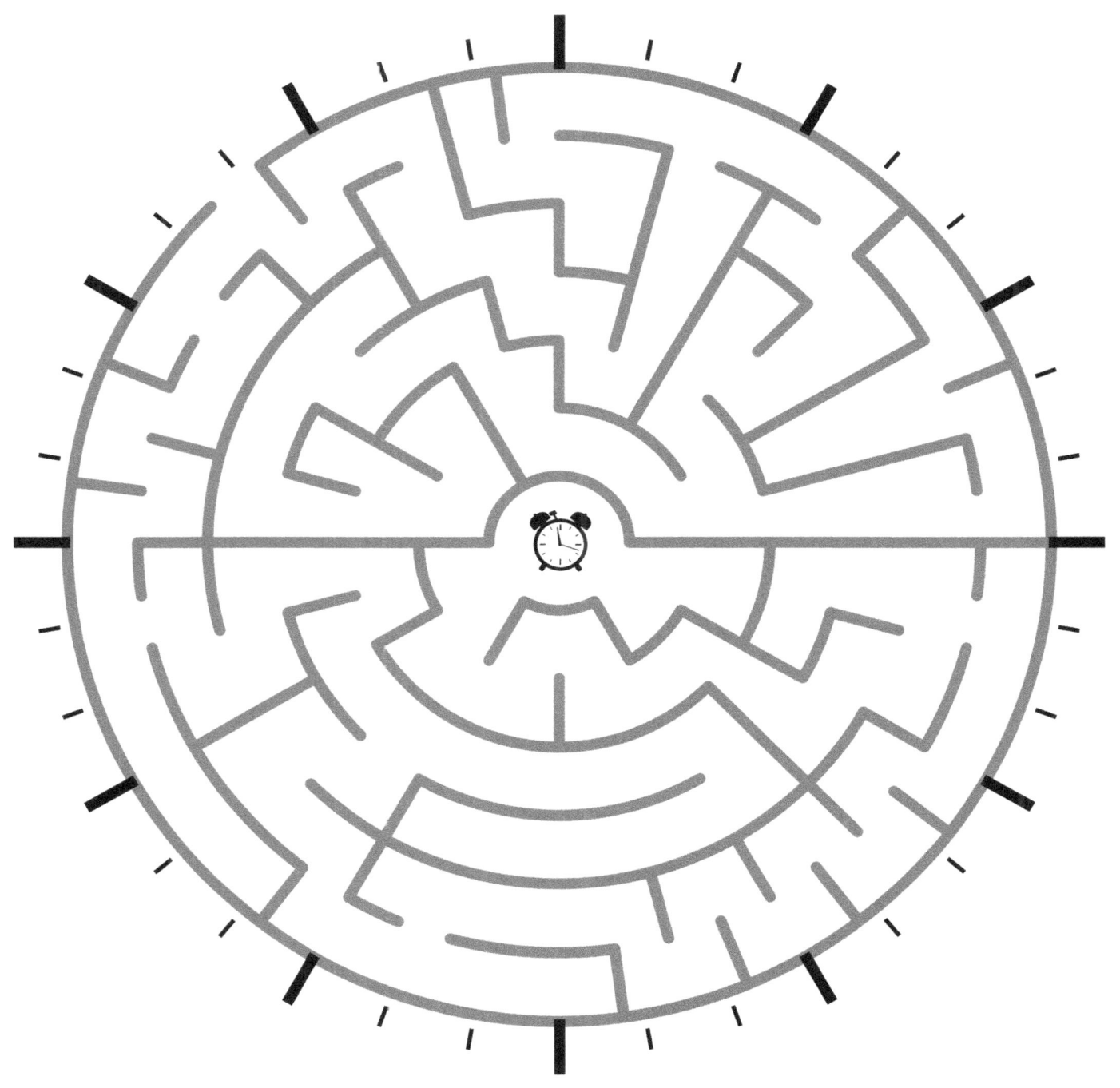

What Time is the Entrance:_____

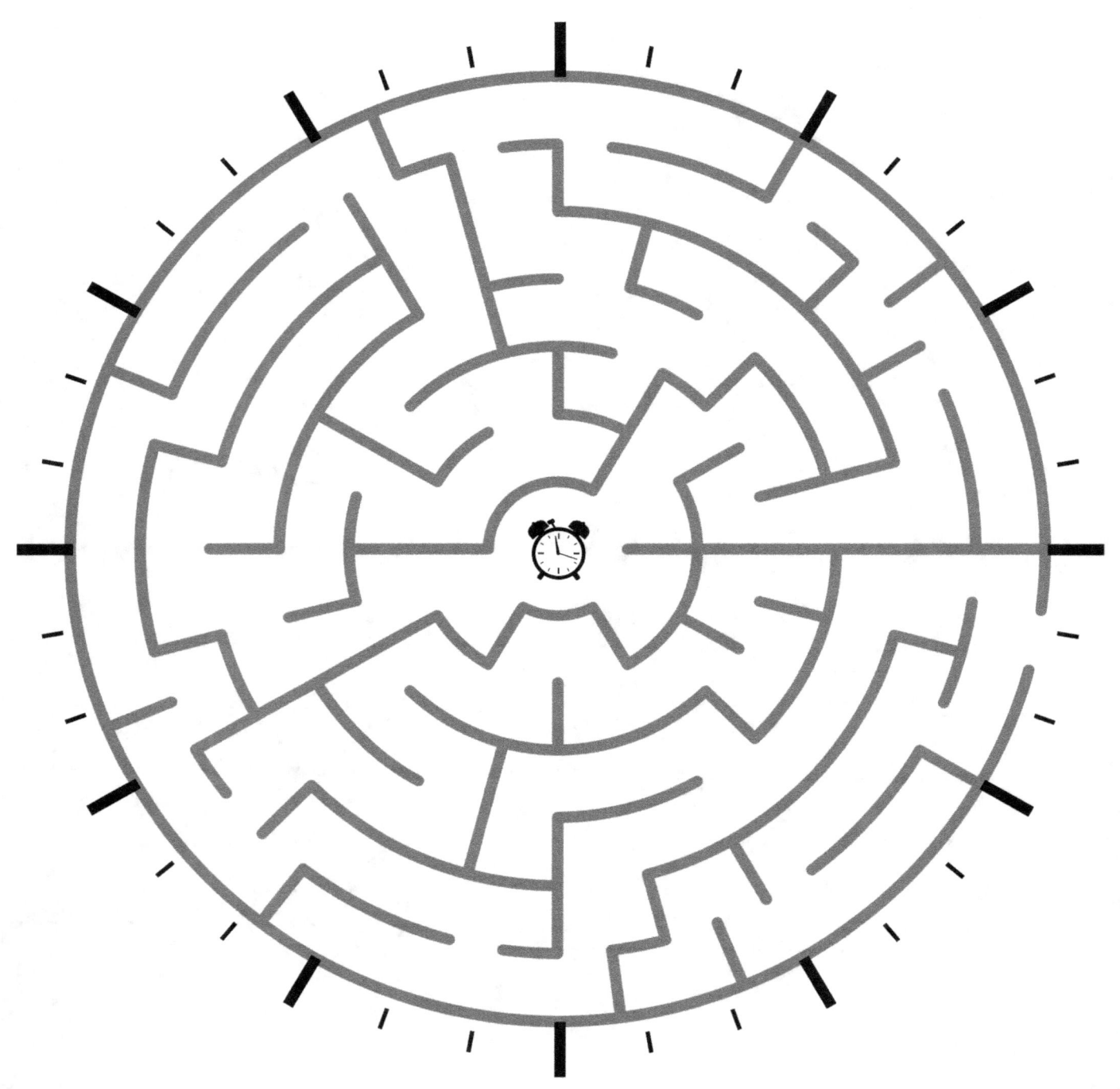

What Time is the Entrance:_____

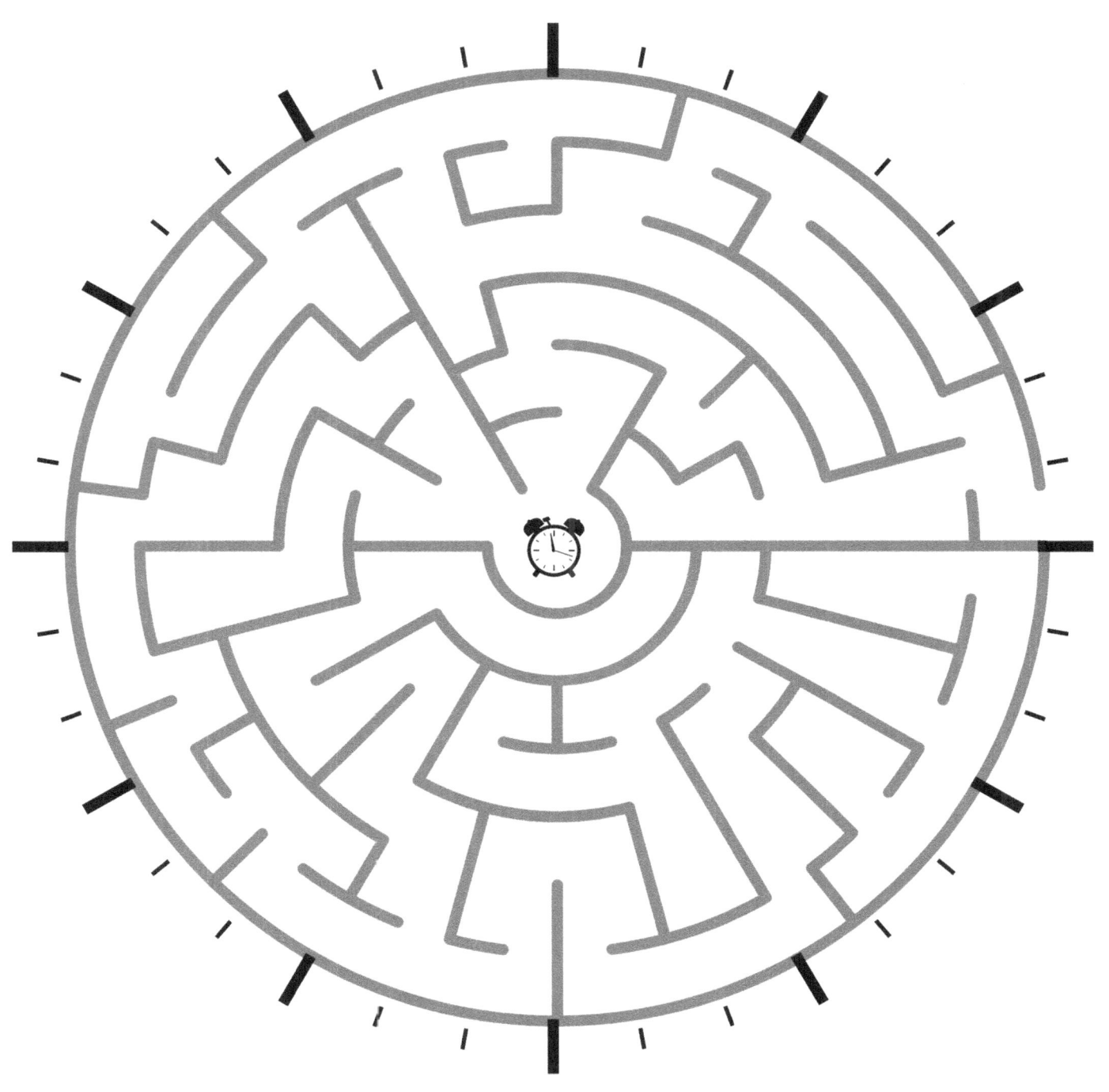

What Time is the Entrance:_____

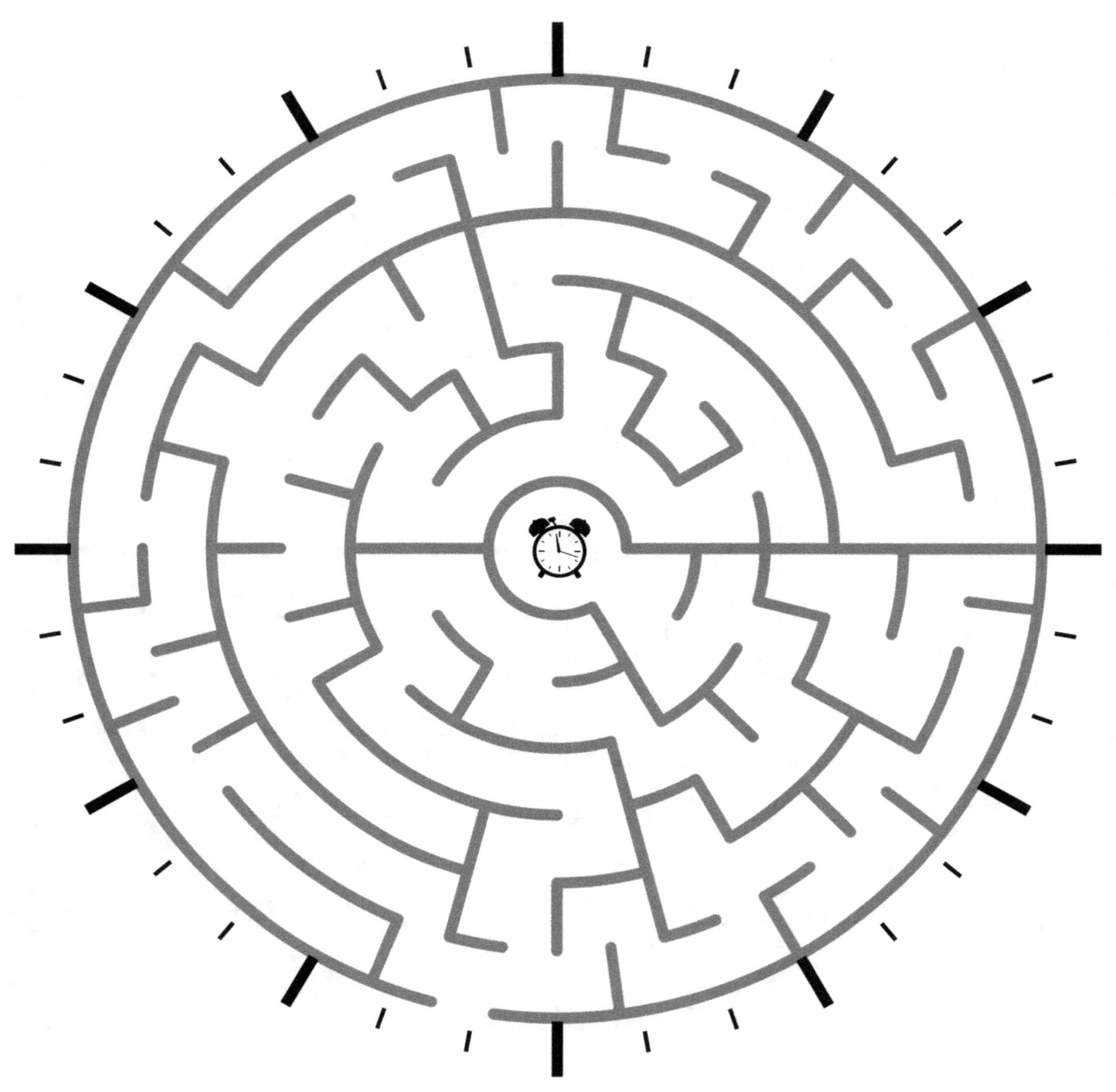

What Time is the Entrance:_____

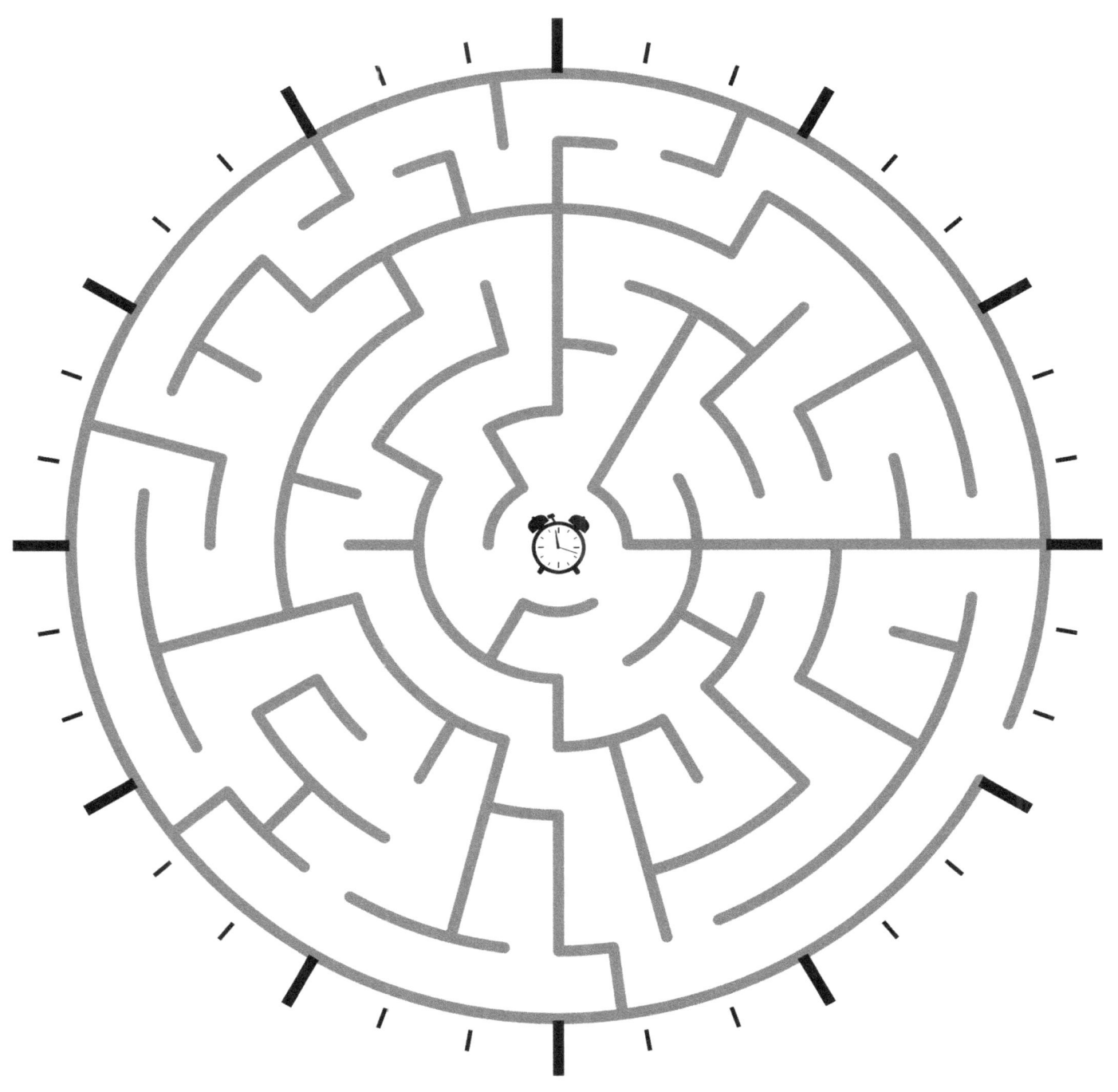

What Time is the Entrance:_____

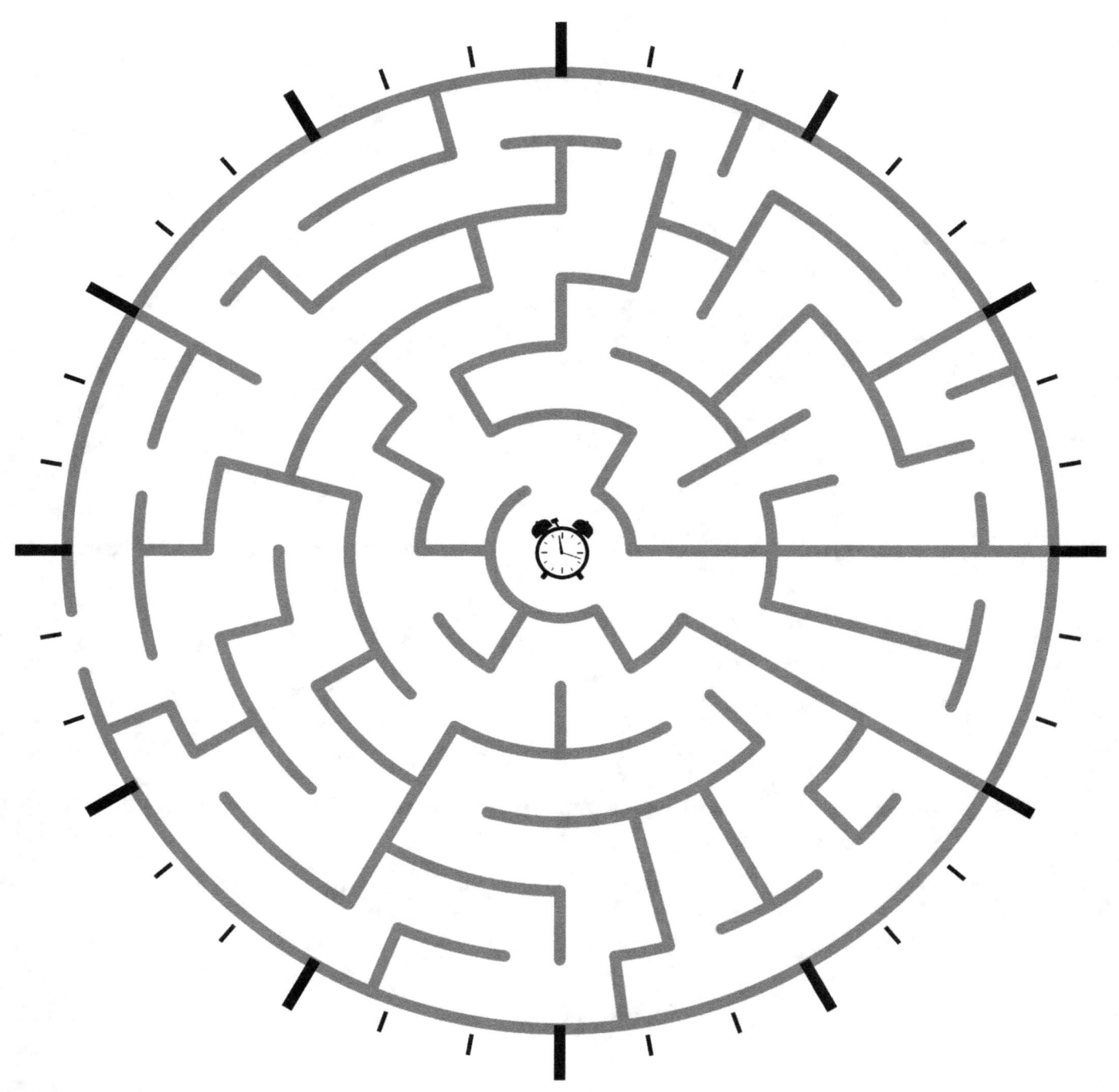

What Time is the Entrance:_____

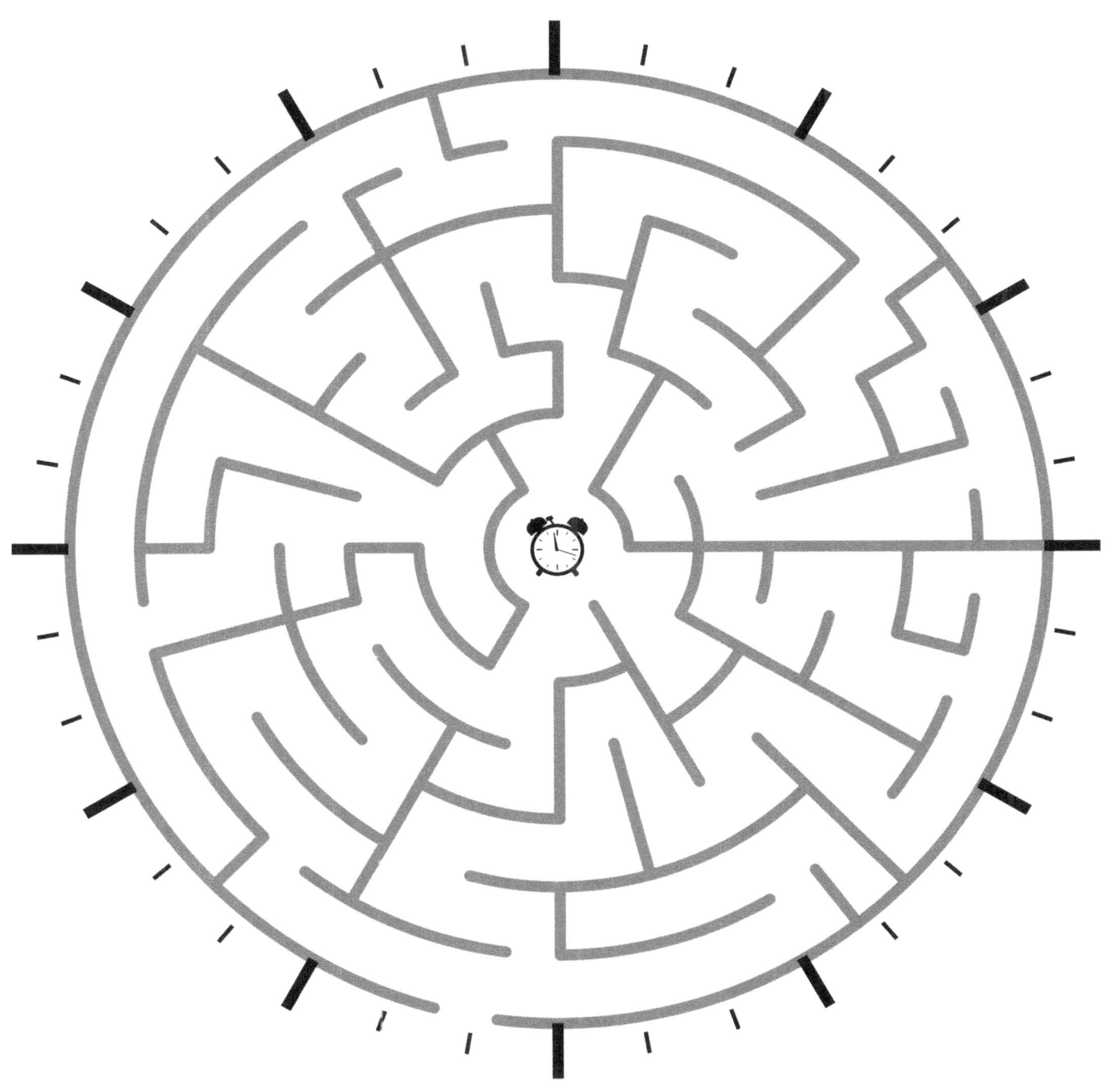

What Time is the Entrance:_____

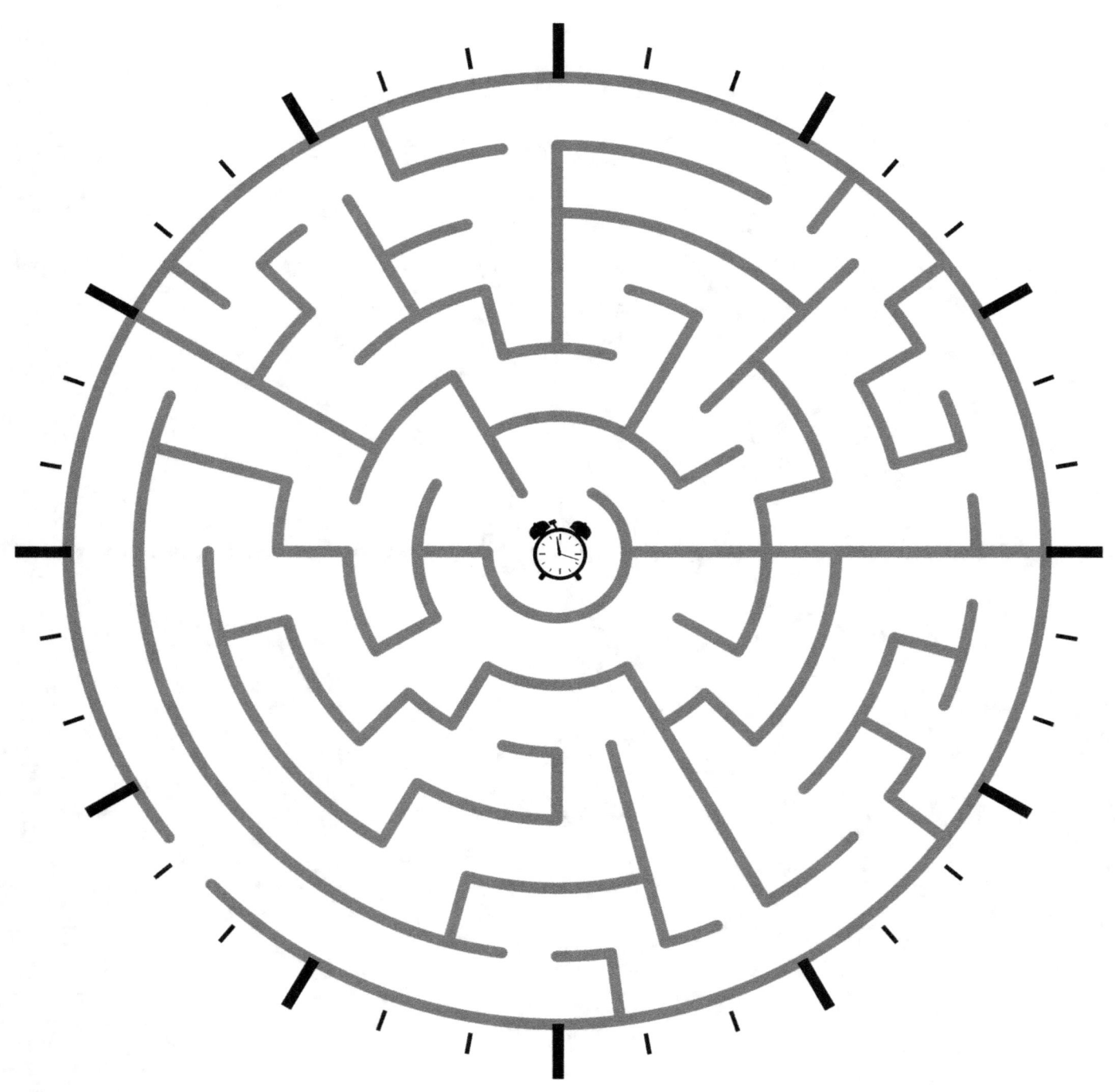

What Time is the Entrance:_____

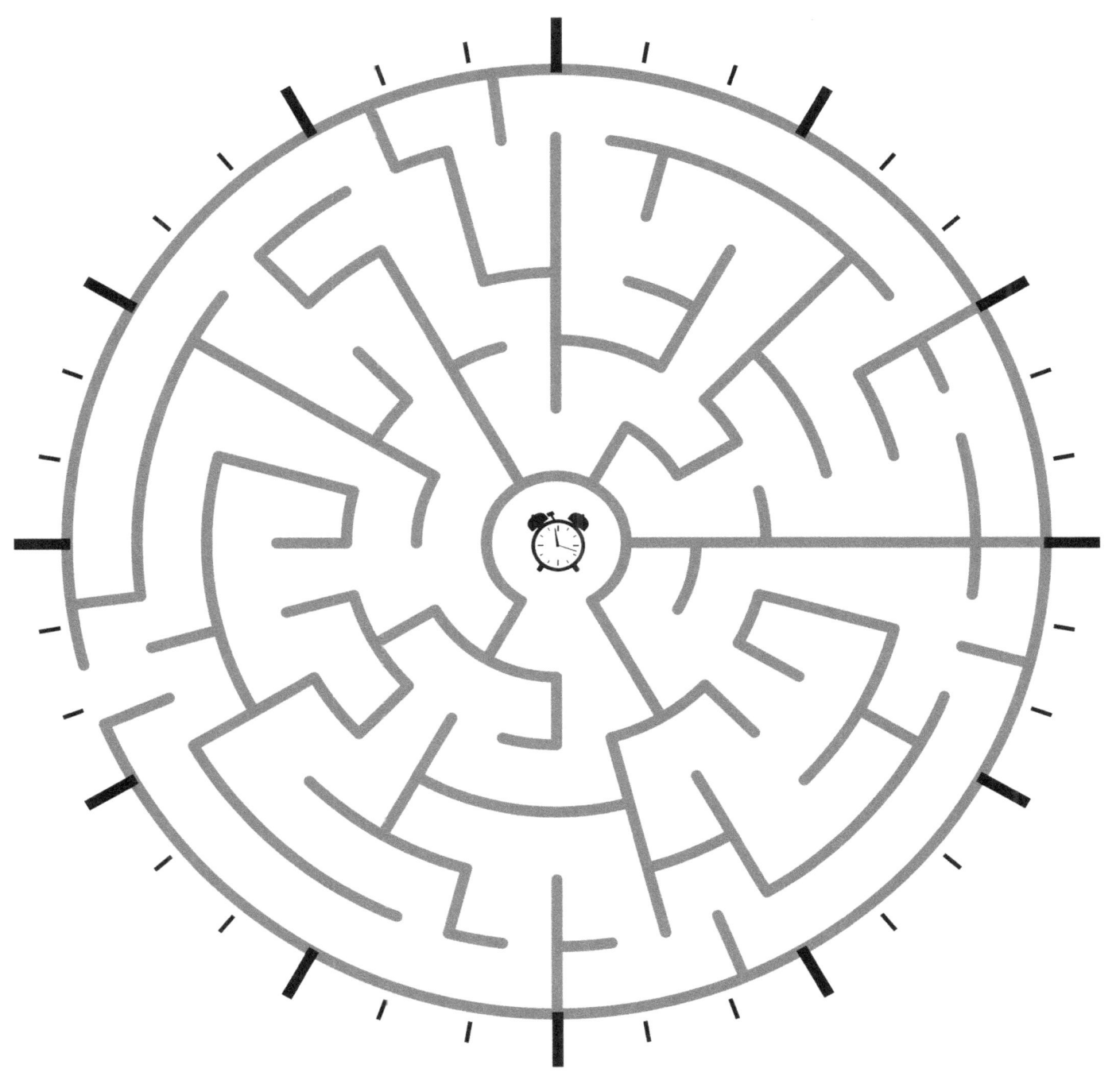

What Time is the Entrance:_____

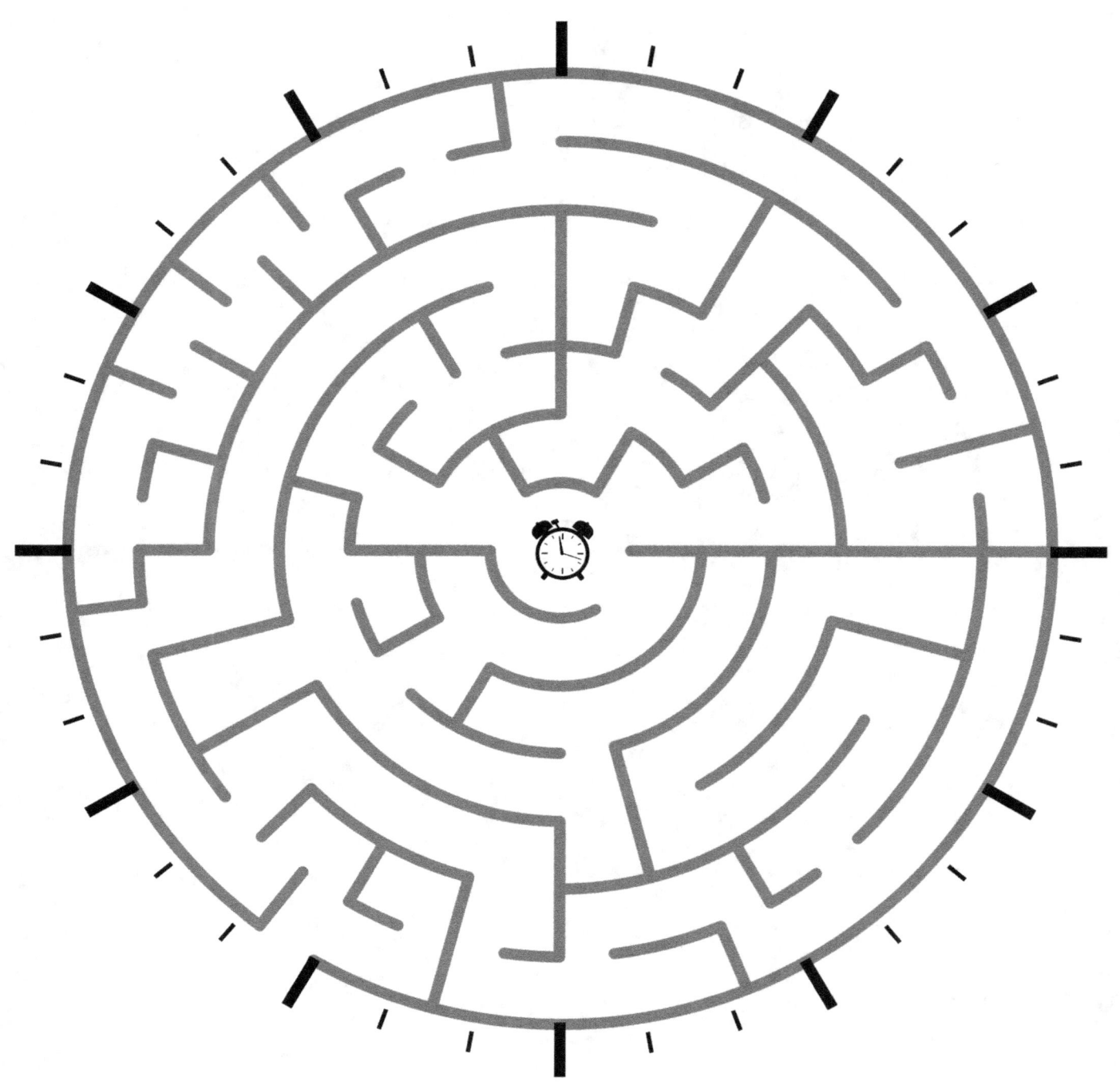

What Time is the Entrance:_____

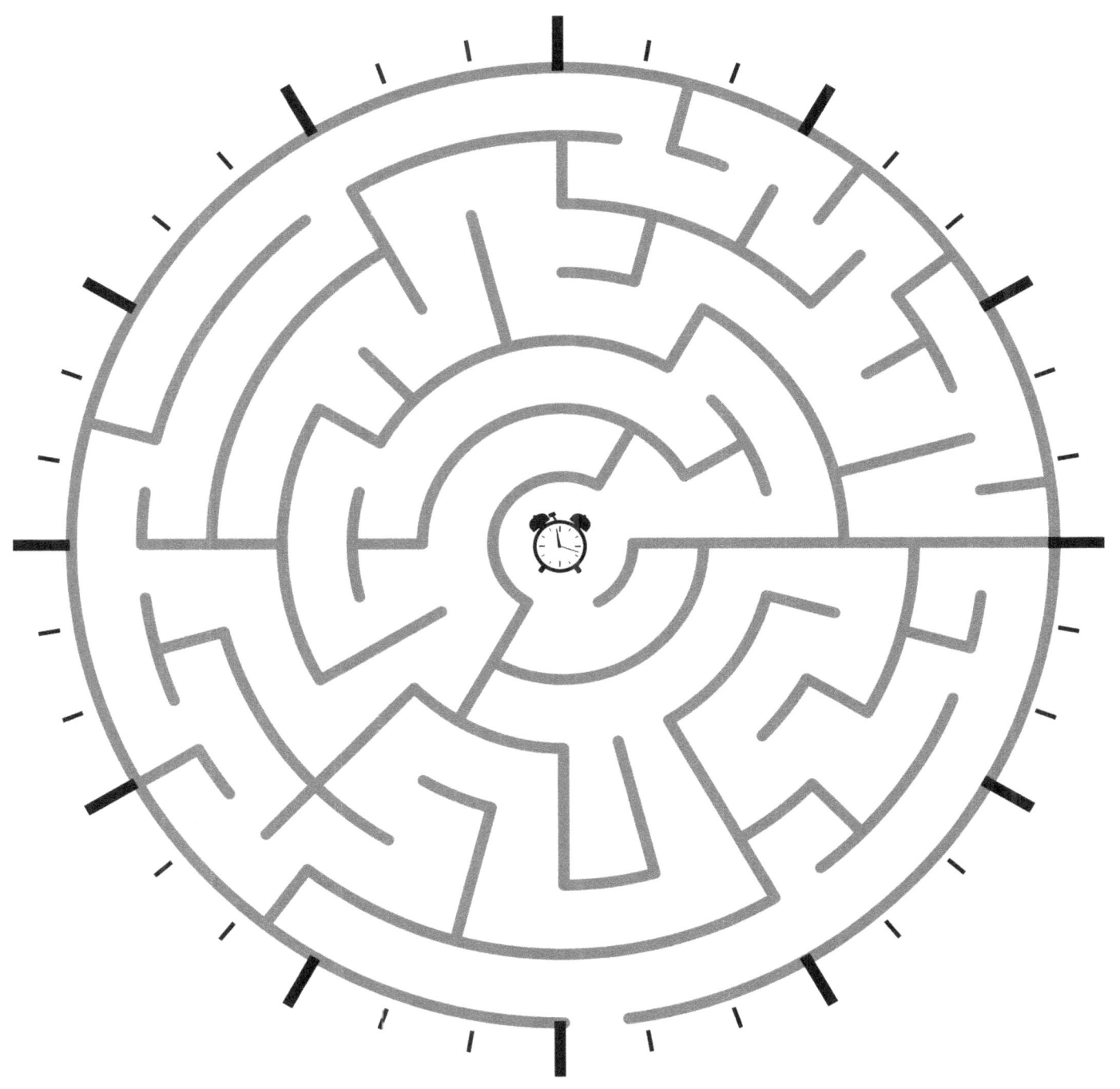

What Time is the Entrance:_____

Solutions

Page 1 - 12:20

Page 2 - 1:40

Page 3 - 12:20

Page 4 - 1:00

Page 5 - 7:20

Page 6 - 1:40

Page 7 - 7:20

Page 8 - 12:20

Page 9 - 11:00

Page 10 - 11:00

Page 11 - 8:00

Page 12 - 1:40

Page 13 - 1:00

Page 14 - 7:20

Page 15 - 4:00

Page 16 - 10:00

Page 17 - 7:20

Page 18 - 2:00

Page 19 - 10:20

Page 20 - 4:00

Page 21 - 4:40

Page 22 - 3:40

Page 23 - 9:00

Page 24 - 3:20

Page 25 - 5:00

Page 26 - 11:40

Page 27 - 4:20

Page 28 - 6:00

Page 29 - 8:00

Page 30 - 8:00

Page 31 - 1:00

Page 32 - 10:20

Page 33 - 9:00

Page 34 - 2:40

Page 35 - 12:00

Page 36 - 9:40

Page 37 - 3:20

Page 38 - 4:40

Page 39 - 11:40

Page 40 - 12:00

Page 41 - 2:00

Page 42 - 8:20

Page 43 - 6:20

Page 44 - 4:00

Page 45 - 2:20

Page 46 - 11:00

Page 47 - 9:00

Page 48 - 10:40

Page 49 - 7:40

Page 50 - 8:40

Page 51 - 5:20

Page 52 - 1:40

Page 53 - 2:20

Page 54 - 7:40

Page 55 - 3:00

Page 56 - 7:40

Page 57 - 5:00

Page 58 - 2:20

Page 59 - 4:00

Page 60 - 5:40

Page 61 - 1:00

Page 62 - 1:00

Page 63 - 7:00

Page 64 - 5:40

Page 65 - 5:40

Page 66 - 5:40

Page 67 - 10:40

Page 68 - 4:00

Page 69 - 4:20

Page 70 - 2:20

Page 71 - 12:20

Page 72 - 8:20

Page 73 - 5:00

Page 74 - 11:20

Page 75 - 7:00

Page 76 - 11:00

Page 77 - 10:40

Page 78 - 12:00

Page 79 - 2:20

Page 80 - 12:20

Page 81 - 7:20

Page 82 - 7:20

Page 83 - 2:20

Page 84 - 9:20

Page 85 - 6:20

Page 86 - 10:40

Page 87 - 3:20

Page 88 - 3:00

Page 89- 6:20

Page 90 - 4:00

Page 91 - 8:40

Page 92 - 6:20

Page 93 - 7:40

Page 94 - 8:20

Page 95 - 7:00

Page 96 - 6:00

Follow us on social media for free content on Benny the Lovable.

Youtube

FaceBook

Instagram

Pinterest

www.ingramcontent.com/pod-product-compliance
Lightning Source LLC
Chambersburg PA
CBHW081538120626
46550CB00009B/2773